SOUTHAMPTON DOCKS

SOUTHAMPTON DOCKS
LOOKING BACK AT BRITAIN'S PREMIER PORT

ANDREW BRITTON

The
History
Press

*Cover illustration*s: *Front*: The departure of *Arcadia* from Southampton to Sydney on 21 August 1954 (Pursey Short/Britton Collection). *Back*: The Cunard RMS *Franconia* and RMS *Carmania* raise steam at 101 Berth (Gwilym Davies/Britton Collection)

First published 2014

The History Press
The Mill, Brimscombe Port
Stroud, Gloucestershire, GL5 2QG
www.thehistorypress.co.uk

British Library Cataloguing in Publication Data.
A catalogue record for this book is available from the British Library.

ISBN 978 0 7524 9881 2

Typesetting and origination by The History Press
Printed in India

CONTENTS

INTRODUCTION

Southampton is a city synonymous with the sea. Southampton Docks is the cornerstone of the city – it is her heritage and her way of life. Since my early childhood days in the 1950s, I can remember the road signs on the boundary of Southampton proudly proclaiming, 'Southampton – The Gateway to the World'. Just a few miles along the road one could peer out of the car window to see the docks and know immediately why Southampton was pleased to boast that title. Approaching Southampton, the visitor would see before him or her a vast array of liners, cargo ships, ferries, coasters, tugs, tankers and shipping which would set the pulse racing. As a child sleeping in my bed at Beaulieu Road station in the New Forest, it was possible to hear the haunting whistles of ocean liners and tug boats in the docks. I would close my eyes and dream about Southampton Docks.

I was privileged to frequently visit the docks, to have almost unlimited access to the great ocean liners and to meet up with family and friends who worked on them. My grandfather was leader of the orchestra on the Cunard White Star Line, while my uncle, Joe Webb, worked on the Cunard RMS *Queen Elizabeth* in the engine boiler rooms and later in the docks. A family friend, Sid Deeming, worked on the P&O cruise ship *Oriana,* and it was always a great pleasure to visit him on board his ship. It was so exciting to peep behind the scenes in the docks, watching steam locomotives shunting long lines of wagons, witnessing the arrival and departure of liners and the preparation for a voyage, and to have the freedom to explore a liner from top to bottom in the company of a family member or friend.

With the passage of time, memories of Southampton Docks came flooding back and I set out to record them, which has led to this book. Featured are many ships of Southampton, ranging from the humblest tug to the mighty Cunard Queens. Also captured are behind-the-scenes shots of dock life and many stunning aerial action views dating back to the bygone age of the four-funnel liners.

Many friends have helped by allowing me to include their original work and several have generously donated this material. I am indebted to Mrs Hilda Short and the estate of the late Pursey Short for the donation of Pursey's colour slide material of Southampton Docks. Pursey's aerial photography of the docks is second to none. Thanks, also, to the estate of the late Gwilym Davies for their support in acquiring the entire collection of maritime slides of this distinguished photographer. I am very grateful to Randy Holmes and the Church of Latter-Day Saints, USA, for allowing me to purchase the entire Arthur Oakman original slide collection for inclusion in this and future books. Additionally, the G.R. Keat and Norman Roberts slide collections have been sold to me for specific inclusion in this book, for which I am extremely grateful. Petroleum giant Esso and the Cunard Line have very generously given me their original slides of Southampton Docks and the liners. Similarly, I am sincerely grateful to Graham Cocks for the generous gift of all his slides to this project. Barry Eagles, Bryan Hicks, John Goss, David Peters, John Cox, Alan A. Jarvis, Tom Hedges, A.E. Bennett and John Wiltshire have also given permission for their original work to be been included and I am very grateful to them.

I owe a great debt of thanks to Jim McFaul of the World Ship Society for his considerable assistance in locating colour slides for this book.

The idea for the ocean liner book series began with my cousin, Jennifer. News soon spread around the family, and Uncle Joe Webb (of RMS *Queen Elizabeth* fame) and Auntie Jean in Southampton began to encourage and arm-twist. Bernard Webb (ex-RMS *Queen Mary*) added his support and the project was on, and cousins Tim and Wendy Webb have helped me enormously. Keith Hamilton, Shipping Correspondent of the *Southampton Echo* newspaper has been very supportive over the years and encouraged me immensely. He has provided so many helpful tips, including selecting a suitable publisher. Thank you, Keith.

The unique colour material is previously unpublished and follows a geographical sequence from the Test estuary down Southampton Water to the Solent.

As always, my friend and neighbour Michael Jakeman has thoroughly checked the content of this book. I would like to express my appreciation for all his time and generous assistance.

I am extremely grateful to my brother-in-law, Mike Pringle, who meticulously scanned each and every slide. This amounts to months of unseen hard work. Without his help, this book and the whole series would not have come into being. I also owe my sister, Ruth, a massive thank you for all her patience and encouragement. A special thank you should go to my wife, Annette, and my sons Jonathan, Mark and Matthew for encouraging me and putting up with masses of colour slides, boxes of ocean liner logbooks and shipping artefacts.

I dedicate this book to my new grandson, Toby.

GEOGRAPHICAL LOCATION

The geographical location of Southampton Docks lies at the northern point of the deep-water estuary of Southampton Water at the confluence of the River Test and River Itchen. Geographically, this is area is defined by Her Majesty's Department for the Environment as a 'ria' in the Hampshire Basin, which was formed at the end of the Ice Age.

The London & South Western Railway identified Southampton Docks as geographically important for the following reasons:

- The shelter afforded by the Isle of Wight protects the port from storms.
- The deep-water channel of Southampton Water provides a sheltered approach to the port and is deep enough for the largest liners.
- Southampton experiences a double high tide, which is due to the sudden narrowing of the English Channel between Portland Bill and the Cherbourg Peninsula. The first high tide reaches Southampton via the Solent and the second high water via the Spithead. The second high water occurs approximately two hours after the first, and it is this prolonged period of deep water which is of such fundamental immense value in manoeuvring very large vessels.
- Southampton possesses an extensive water frontage owing to the fact that it's situated on a triangular peninsula between the rivers Test and Itchen.

◄ The Union-Castle Line *Windsor Castle* is seen making a high-tide dash west out of Southampton past the Needles on the Isle of Wight. (Norman Roberts/Britton Collection)

POTTED HISTORY

EARLY HISTORY

The Port of Southampton can trace its historical origins back to Roman times when it was settled and known as Clausentum, around AD 70. The Romans established the port to serve the settlements at Winchester and Salisbury, but there is evidence that it was abandoned when the Romans departed around AD 407.

Approximately 300 years later, the Saxons built a new settlement known as Hamwick (later Hamtun) along the banks of the River Itchen. The population increased to an estimated 5,000 inhabitants, and archaeologists have discovered evidence of international trade, with imports of wine and pottery from France, Spain, Greece and the Middle East, and exports of wool.

The Viking king Canute was crowned king of England in Southampton in 1016 and his twenty-year reign was peaceful and uneventful. Viking legend has it that while in Southampton he sat on the shingle shoreline on his throne and commanded the waves of the incoming tide to stop and not wet his leather boots; the incoming Southampton Water tide naturally ignored him. Canute was attempting to show his courtiers that he was a mere mortal and they should worship God instead.

The threat to the Port of Southampton at this time came from the North Sea, from menacing Viking longships, and as a result Saxon settlement began to decline. However, following the Norman invasion of 1066, large numbers of Norman immigrants began to settle and the port began to establish itself as a departure point for English armies and supply ships on their way to France.

During the Hundred Years' War, the rudiments of a shipbuilding industry began constructing naval ships. Perhaps the most notable vessel from this period, built in 1418 by William Soper, was HMS *Grace Dieu*, the flagship of King Henry V. She was constructed in Watergate Quay, which is now occupied by Town Quay. Built with local timber from the New Forest, the HMS *Grace Dieu* was more than 200ft long with a displacement of 2,750 tons. She was later destroyed by fire on the River Hamble just off Southampton Water in 1439.

Over the next century, the Port of Southampton became more established with a flourishing import trade from Genoa and Venice of silk, spices, perfumes, alum and wood (used in the wool-dyeing process). Many languages could be heard in the port and some visitors stayed to set up trading centres. An historic building that survives from this period is the old Wool House, which is situated across Canute Road from the Royal Pier. It was originally constructed as a warehouse and was until recently the maritime museum.

In 1620, the Pilgrim Fathers sailed from Southampton for North America on the *Mayflower* and the *Speedwell*. The *Speedwell* had sailed from Holland to meet up with the *Mayflower*. As the two vessels sailed west along the south coast, *Speedwell* was deemed too leaky and put into Plymouth for repairs. Her personnel and stores were transferred to the *Mayflower*, which completed the voyage alone.

During the sixteenth and seventeenth centuries, the Port of Southampton declined as the Italian trade dwindled, with the Port of London capturing much of the market. By 1700, Southampton's international commerce had significantly declined and it was not until 18 April 1803 that measures were put in place to improve, expand and develop the quay facilities with the formation of the Harbour Board. One of their first actions was to enlarge the Town Quay with new walls and, in 1810, to sell the West Quay, which was subsequently demolished. Next, the new Harbour Board had a new pier designed and constructed for passenger services to the Isle of Wight and Channel Islands. It was opened by HRH Princess Victoria on 8 July 1833. Initially this pier was named the Royal Victoria Pier, but this was shortened to Royal Pier, by which name it has been known ever since.

The distinguished historian and *Southampton Echo* newspaper Shipping and Maritime Correspondent, Keith Hamilton, identifies the date of the birth of the Port of Southampton to be 16 August 1836, when a group of businessmen met in the unlikely setting of the George and Vulture Tavern in Lombard Street, London, to plan Southampton Docks. Just a few months earlier, Parliament had passed an Act allowing the Southampton Dock Company to construct the first real berths following the acquisition of 216 acres of mudflats close to Town Quay at a nominal cost of £5,000. The news of this development was greeted with strong opposition in some quarters in Southampton. Heated debates took place over whether '… a set of unknown and irresponsible individuals could turn so great an extent of land to beneficial purposes'.

The Southampton Dock Company pressed on with its scheme and on 12 October 1838, the same year as the coronation of Queen Victoria, a crowd of 20,000 witnessed Rear Admiral Sir Lucius Curtis lay the foundation stone of Southampton Docks.

The siting of the foundation stone was originally the dock's Entrance Gate Number 1 at the Itchen end of Canute Road. In 1900, the foundation stone was excavated and incorporated into the boundary wall. In 1948, the foundation stone was recovered and removed to a permanent monument near Central Road, near the north end of Ocean Dock. More recently it has been fully restored for future preservation.

Perhaps the coming of the railway to Southampton was a major incentive to develop the Southampton Docks. A station was first authorised on 25 July 1834 for the London & Southampton Railway (which was later the LSWR, London & South Western Railway). The station opened on 10 June 1839 at Southampton, but it was not officially operational until 11 May 1840, as the track had not been completed at Winchester. A superb terminal station building was built in 1839 to the design of Sir William Tite. To this was added the impressive South Western Hotel, which dwarfed the station. The running line was extended into Ocean Dock Terminal, thus allowing through boat trains from London to terminate on the quayside. As the first steam-hauled trains arrived, carrying their excited passengers to the ships, a signal was sent for the rapid expansion of the port and the development of facilities, which attracted massive population growth.

A huge boost to the emerging business came from James McQueen of the Royal Mail Steam Packet Company, who announced that Southampton Docks had been selected as their base to operate its ships to the Caribbean as well as the Americas.

Construction progress on the dock development was so advanced that by 29 August 1842, the first ships, P&O Line's *Tagus* and *Liverpool*, could be accommodated. As the railway was in place it was possible to convey passengers from quayside to the capital. 'Success breeds success', as the saying goes, and it soon became apparent that the Outer Dock, measuring 16 acres with the provision of 2,620ft, could not meet the demand. Consequently, plans were put in place for further facilities and for three dry docks.

By 1854, the three new dry docks had been commissioned, the largest of which measured 400ft by 21ft. The first ship to use the new dry-dock facility was the Royal Mail Company's *Forth* in July 1846, to undertake bottom coppering and deck renewal. Meanwhile, a new Inner Dock was pressed into service, still incomplete, in 1851. It had an area of 10 acres with 2,575ft of quays and a double-lock gated entrance of 46ft wide. A curious feature of the Inner Dock was the coal bunker shed at its landward side. This was provided for the early steam ships' bunkering facilities and was famed for being a dark and gloomy location shrouded by black clouds of coal dust.

In 1854, at the time of the Crimean War, the Inner and Outer Docks were thrust into handling the bulk of military transportation. Many ships were requisitioned to convey regiments of troops to Sebastopol. P&O are recorded as conveying over 100,000 men and 20,000 horses to this conflict. Southampton Docks were to witness repeated military use in the future Boer War, the First and Second World Wars, and more recently the Falklands War.

The increasing size of steam ships rendered the Inner Dock totally inadequate for anything but the smallest vessels. Enlargement and deepening plans were drawn up and implemented by 20 May 1859, specifically to accommodate the P&O *Pera*. The Victorian era saw the erection of the Itchen Quays, completed in 1876, and the construction of a fourth dry-dock facility. Further massive investment was required to expand and construct new planned docks. However, the Dock Company was finding it difficult to raise sufficient capital to finance the new works. With the approval of Parliament, the Dock Company entered into a loan agreement with the London & South Western Railway for the sum of £250,000. This enabled the construction of the new deep-water Empress Dock, which was opened by Her Majesty Queen Victoria on 26 July 1890.

▲ The traffic is halted at Canute Road crossing for S15 No 30839 to run into the docks to collect fitted freight for London. (Alan A. Jarvis)

The Empress Dock had a water area of 18½ acres and allowed the largest ships to arrive or sail at any state of the tide.

Moreover, incorporated into the new Empress Dock, on 3 August 1895, was the largest dry dock in the world at this time: The impressive Prince of Wales Dry Dock measured 745ft in length. Sadly, it was closed in 1976 and filled in a year later during the redevelopment of the Eastern Docks.

The Empress Dock would become the centre of Southampton's fruit trade, with fast-fitted freight trains steaming out day and night to supply London's Covent Garden and the Midlands' wholesale fruit markets.

In the years that followed, the port facilities continued to expand, but a highly significant move by the American Line to transfer their New York mail service from Liverpool to Southampton established the Hampshire port as the centre of transatlantic trade. In 1907, Southampton claimed another of Liverpool's major customers with the transfer of the White Star Line's North Atlantic service to the South Coast.

Anticipating the potential growth in North Atlantic traffic, a new dry dock was planned. This was the Trafalgar Dry Dock, named after Nelson's famous victory over the French and, fittingly, opened on on 21 October 1905 – Trafalgar Day. It far exceeded the size of earlier dry docks, with dimensions of 912ft long and 100ft wide, with a capacity of 100,000 tons of water. It was considered by the builders to be the last word in dry docking and capable of handling all future needs, yet within six years of opening it could barely cater for the developing liners of the day. Cunard Line's *Berengaria* could only just squeeze into the Trafalgar Dry Dock with 10in to spare either side of her stern.

The next landmark date in the history of Southampton Docks came in 1911 with the opening of the famous White Star Dock, renamed Ocean Dock in 1922. It signalled that Southampton was now the 'Gateway to the World' and Britain's premier passenger port. The first ship to sail from the new White Star Dock was the 46,000-ton *Olympic* in 1911.

A year after the opening of the White Star Dock, perhaps the most famous ocean liner in the world pulled away at the start of her maiden voyage on 10 April 1912 – none other than the RMS *Titanic*. After sailing down Southampton Water and across the Solent, *Titanic* called at Cherbourg in France and Cobh in Ireland before sailing west out into the Atlantic heading for New York. Four days into the crossing, about 375 miles south of Newfoundland, she hit an iceberg at 11.40 p.m. ship's time. This caused *Titanic's* plates to buckle inwards along her starboard side opening up five of her sixteen watertight compartments, causing the unsinkable liner to disappear beneath the waves. Just two hours after the *Titanic* sank, the RMS *Carpathia* arrived at the scene where 705 survivors were rescued.

The disaster made headlines across the world, but nowhere was the tragedy of the disaster more felt than in Southampton, where more than 500 households lost at least one family member. The sinking of the *Titanic* claimed more than 1,500 lives.

To the Britton household, and the author's grandfather, Alfred Britton, who lived in Southampton, it was a very lucky escape from the jaws of death. It has often been recalled within the family that back in 1912 the author's grandparents watched the liner slowly sail up Southampton Water after completing her sea trials, watched by cheering flag-waving crowds. As the gleaming 46,329-ton *Titanic*, under the command of Captain Edward Smith,

laboriously turned to edge her way into the White Star Dock, the author's grandparents observed her suck in a neighbouring vessel and collide with it as she displaced the water in the dock. After safely docking the 852ft liner, the filthy task of bunkering coal commenced to fuel her boilers. Alfred Britton recalled to the family that the coal supplies were pilfered from neighbouring ships, there being an industrial dispute at the time.

Across Southampton, the air was filled with excitement, and Alfred Britton carefully packed his violin to join the maiden voyage of the doomed liner. Alfred had been leading violinist and bandmaster on the Royal Mail and White Star Lines and was looking forward to serving on the already famous *Titanic*. That night, the author's grandmother had a chilling premonition and when she awoke in a cold sweat she begged Alfred not to sail with the ship. With some regret, the following day Alfred Britton watched *Titanic* sail, fearing he had missed out on a great opportunity – and plenty of tips, since the liner had a full complement of rich passengers.

The Britton family recalled that when news of the sinking of *Titanic* reached Southampton there was a feeling of disbelief, shock and horror. Along the streets of Northam in Southampton, curtains were closed as a mark of respect for deceased family members. There was much weeping behind the scenes. The author's father often said the scarring effect of guilt and fear of ostracism could well have led Grandfather Alfred to take to the bottle! Without that premonition there isn't the slightest doubt that Alfred would have gone down leading the orchestra on *Titanic* as it played, 'Nearer my God to Thee.' He was that sort of man.

In 1913, the management structure of Southampton Docks was reorganised. Southampton's Harbour Board now included representatives from the Admiralty, War Office, Trinity House, The Board of Trade, London & South Western Railway and Southampton Chamber of Commerce.

3

THE FIRST WORLD WAR 1914–18

As in peace, so in war: the Port of Southampton has played a leading part in Britain's wavering fortunes, and was crucial during the two world wars. The strategic geographical location and ability to deal efficiently with shipping and associated traffic was quickly recognised in 1914 at the outbreak of the First World War. It almost immediately became Britain's number-one port, under government control, and acted as the main point of departure for troops leaving the home shores to fight in the trenches of the Western Front. Southampton Docks was also the main port for the arrival of colonial troops and, after 1917, American GIs.

By midday on Wednesday 5 August 1914, the army general headquarters had arrived in Southampton and was stationed at the Polygon Hotel prior to sailing for France on the SS *Ermine* on 14 August. All normal traffic in the docks was suspended until further notice to make way for military activities. Her Majesty's Home Office First World War records state that on 22 August 1914 a total of 536 officers, 16,364 other ranks, 4,572 horses, 17 tons of hay and fodder, 72 artillery pieces, 690 vehicles and 260 bicycles were loaded for sailing from Southampton. Arriving in large numbers at the Port of Southampton were Belgian refugees who were taken to Tower House in Welbeck Avenue

▲ A view of the Old Eastern Docks in the early twentieth century. (Britton Collection)

for temporary quartering. Captured German prisoners of war were escorted off the ships and marched up the Western Esplanade to the Shirley skating rink where they were dispatched via St Deny's railway station to POW camps. Soon the sight of wounded British and Empire troops became an all too familiar sight, some blinded by mustard gas or with shell shock from the trench warfare. Nursing staff and doctors greeted the wounded at the quayside and escorted them to the Military Clearing Hospital at Eastleigh.

Many of the illustrious passenger liners that had frequented the Port of Southampton were requisitioned by the British Government as troopships, or converted to armed cruisers and hospital ships. The third of the *Olympic* four-funnel trio, the *Britannic,* made her maiden entry into Southampton in December 1915 after being hurriedly completed for service as a hospital ship. According to members of the Britton family, the *Britannic* was very reminiscent in appearance to the ill-fated *Titanic.* After entering Southampton, she was bunkered and packed with food and supplies ready to set sail at 2.23 p.m. on 12 November 1916 on her sixth voyage to the Mediterranean Sea in support of the Dardanelles offensive. Nine days later, in the Aegean Sea, a large explosion on the starboard side shook the ship; whether it was an enemy torpedo or mine is not certain. This resulted in *Britannic* sinking with the loss of thirty lives. Strangely enough, at the time of writing, an oak door and port hole recovered from the wreck are on sale at a second-hand shop in Northam, Southampton.

One hidden secret gem of the history of Southampton Docks during this period concerned the installation at Shed 50 within the docks. Shed 50 was operated as a secret works and stores for the refurbishment of guns damaged at the front. It was also used as a temporary store for ordinance – if this had exploded, it could have had devastating implications. A military railway was also constructed between the west station, along Western Esplanade, to a jetty for direct loading of troop trains onto special train ferries.

During the 1914–18 war, over 8,149,685 officers and men passed through the port. This averages to a daily total of over 4,000 men. A further 859,830 horses and pack mules, 15,266 guns, 179,069 vehicles and 1,983 aircraft were transported through the docks.

The chairman of the London & South Western Railway, H.W. Drummond, wrote: 'It is difficult to imagine how the war could have been carried on and brought to a successful conclusion had it not been for the magnificent facilities available at the Port of Southampton.'

4

THE INTERWAR YEARS

A general feeling of great optimism and potential expansion of the Port of Southampton filled the air at the time of the Armistice. The year 1919 saw Cunard switch its express service from Liverpool to Southampton. The following year the Canadian Pacific Steamship Company inaugurated a Southampton to Quebec service which introduced their distinctive 'White Empress' liners to the local skyline. On 29 March 1922 the White Star Line *Majestic* – then the largest liner in the world – arrived at Southampton. Former Hamburg America Line's *Bismark,* she was acquired as a war reparation. This was the last liner in the jigsaw puzzle for a twice-weekly transatlantic service comprising the Cunard liners *Aquitania, Mauretania* and *Berengaria* working alongside White Star Line's *Olympic* and *Homeric.* (The *Berengaria* and *Homeric* were also German war reparations, being the former *Imperator* and *Columbus.*)

With the grouping of the railway companies in 1923, Southampton Docks passed into the hands of the Southern Railway. The newly formed company immediately embarked on a reclamation scheme of 400 acres of tidal mudflats along the northern foreshore of the River Test estuary. The Southern Railway also identified an acute requirement for new dry-docking facilities, which was highlighted by the limitations of the Trafalgar Dry Dock for the new and larger modern liners. The novel temporary solution to this problem was to purchase from Armstrong Whitworth of Newcastle upon Tyne a floating dry dock with a lifting capacity of 60,000 tons. The 960ft long by 170ft wide floating dry dock was delivered to Southampton in April 1924 and opened by HRH The Prince of Wales on 27 June 1924. This floating dry dock presented a bewildering sight

> Liners abound for New York, Cape Town, Buenos Aries and Australia are shown berthed at piers in Southampton in this aerial picture taken in the 1920s. (Associated Press)

to local people and visitors as they witnessed the sight of gigantic liners which were elevated completely out of the water for refits. To use this floating dry dock, seawater would be allowed into its internal tanks to partially submerge it. An ocean liner was then towed in and water would be pumped out. In so doing, the dock was raised in height, taking the ship out of the water and exposing the hull for repair and maintenance.

The plans to create a new set of docks between Millbrook Point and the Royal Pier, creating 1½ miles of deep-water berths, were now being

championed by Sir Herbert Walker, the chairman of the Southern Railway. His contribution to this facility was later acknowledged by naming the main road through the Western Docks after him. The secure foundations to the new Western Docks were built upon 146 concrete and steel monoliths, each 45sq. ft. Intensive labour with round-the-clock shiftwork enabled rapid progress with the project, allowing the first completed section to be opened on 19 October 1932, with the RMS *Mauretania* being the first to dock.

At the far end of the new development, two new graving dry docks were planned for refitting the new 1,000ft-long transatlantic liners. The drainage and excavation of the mudflats was said to be very challenging at this point, and subsequently it was decided to construct only one of the graving dry docks. The dry dock involved the excavation of 2 million tons of soil and chalk. Approximately 0.75 million tons of reinforced concrete was provided to fabricate the walls and basin of the dry dock. It was equipped with four 54in centrifugal pumps to extract 260,000 tons of water in four hours. This graving dry dock was the largest in the world and was opened by King George V (KGV) on 26 July 1933, and suitably named the King George V Graving Dry Dock. The first ship to use the new facility was the White Star Line *Majestic* in 1934.

The total cost of the new Western Docks and the King George V Graving Dry Dock was £10 million, and when complete the Port of Southampton was ready for what has become known as the 'golden age of liners'.

In 1935, the prestigious French Line CGT (Compagnie Générale Transatlantique) transferred their operations from Plymouth to Southampton as the port of call for their liners sailing on the North Atlantic and West Indies services. Enter arguably the greatest ocean liner ever to grace Southampton

◄ Nautical floodlights illuminate the floating dry dock as workmen strive to complete the refit of the *Berengaria* on 4 February 1934. (Associated Press)

Waters: the SS *Normandie*. She entered service in 1935 as the largest and fastest passenger ship afloat. Her stunning design and lavish art deco interiors impressed many. The author's father and grandfather visited the *Normandie* in Southampton for a tour of the great liner and were treated with the finest French cuisine of pastries and cheeses. This left a favourable, lasting impression, but when asked if they could inspect the engine rooms, a swift reply was received: '*Non! C'est interdit!*' She made 139 westbound transatlantic crossings from her home port of Le Havre to New York. The pride of the CGT, the SS *Normandie* was to capture the Blue Riband for the fastest transatlantic crossing at several points in her career and was the main rival to the RMS *Queen Mary* in the interwar years.

A curious accident occurred on 22 June 1936 as the *Normandie* was off Cowes, when an RAF Blackburn Baffin seaplane from Calshot Air Station buzzed the liner on the starboard side. The seaplane collided with the French liner and struck a derrick, landing on the foredeck. This caused damage to a limousine. Luckily there were no injuries – except for a bit of damaged pride. When the *Normandie* sailed later that day, the uninvited guests, both pilot and plane, had to sail with the liner to Le Havre.

The author's father had the pleasure of rowing his dinghy around the German Kriegsmarine cruiser *Admiral Graf Spee* when she visited Southampton in May 1937. The Deutschland Class *Admiral Graf Spee* was to later be deployed to the South Atlantic in the weeks prior to the outbreak of the Second World War, to be positioned in the merchant sea lanes ready for when war was declared. She engaged the Royal Navy in the Battle of the River Plate after sinking nine ships. In a famous incident the *Admiral Graf Spee* was eventually scuttled. During her visit to Southampton, curious young visitors were not encouraged to row around this German war machine, and the irate crew screamed at them to go away.

It was not just ocean liners that attracted the attention of sightseers, for from 1919, flying boats became a familiar sight on Southampton Water. At that time, suitable runways for large aircraft were scarce and engines were unreliable. Consequently, it was felt safer for an aircraft to be fitted with a marine hull rather than wheels and a light flimsy frame, in case it was required to ditch over water if it encountered problems. The first service flew from the Royal Pier to Bournemouth, Portsmouth and the Isle of Wight. During the 1920s, flying boat services were expanded and flights from Woolston to Northern France proved popular. As aircraft technology and reliability improved, Imperial Airways introduced new services from 1934 on their Empire routes to the Mediterranean, Africa and the Far East. The popular flying boats were maintained at Hythe, and a terminal was built at 101 Berth in the new Western Docks. In 1938, passenger operations transferred to 108 Berth.

The proximity of flying boats taking off and landing from the busy Southampton Water did, however, have dangerous disadvantages. In July 1939, the Red Funnel paddle steamer PS *Gracie Fields* was sailing from Cowes and had the top of her mast clipped by a flying boat attempting to take off, which lost its right wing as it hit the paddle steamer's port bow and crashed. Luckily no one was injured. There are records of several, 'heavy landings' of flying boats including one just skimming the Red Funnel steam tug *Calshot* in 1937, in what would be categorised as a near miss by today's aviation standards. During the 1920s and '30s, the Harbour Board exercised control over the flying boats under the Air Navigation Act 1919.

Mention must be made of the Supermarine Works at Woolston, which are known for the success in the Schneider Trophy races. Through the experience gained in the three wins in 1927, 1929 and 1931 with seaplanes, the chief aircraft designer R.J. Mitchell of Supermarine went on to design the Second World War ace fighter plane, the Supermarine Spitfire.

5

THE SECOND WORLD WAR 1939–45

Few could have foreseen that just a few years after the completion of the New Docks Southampton would be thrust into the front line of another world war. Between 1939 and 1945 the Port of Southampton would once again become a focal point for troop activity. Just days before the outbreak of war on 3 September 1939, the docks witnessed a build-up of troops and equipment. A procession of Matilda tanks arrived at the port for loading from Bovington Camp in Dorset and an endless convoy of AEC (Associated Equipment Company) Matador, Scammell and Bedford army lorries, and 6lb anti-tank and 25lb artillery pieces poured into the port from Salisbury Plain for embarkation on requisitioned ships.

Once again, the merchant navy vessels were requisitioned by the Admiralty and some of the larger liners were dispatched to New York, Singapore and Halifax. In 1940, the large floating dry dock was transferred to Portsmouth. Anti-aircraft batteries began to appear around the port along with many observation lookout points. Security around the whole of Southampton Docks was strengthened and dock police were accompanied by armed military guards in case of fifth columnists, who had made their presence felt in the fall of the port of Rotterdam.

With the fall of France in 1940, the nature of the Port of Southampton changed from a supply bridgehead to a vulnerable frontline location under the threat of attack and invasion. The port was closed and further measures were taken to strengthen defences.

Looking from Weston Shore on 18 May 1940, the author's father saw a small armada of inshore cabin cruisers, fishing vessels and barges making their way slowly up Southampton Water. Their decks appeared to be heaving with silent figures – large and small. That night, the author's grandfather informed the family that the pathetic armada was full of refugee families from France escaping the advance of the German armies. Having almost nothing, the stevedores and resident ship crews had a whip-round for the children to buy some sweets and basic essentials. They had been unloaded at the Royal Pier and provided with blankets and hot soup before being taken to the Guildhall and Sportsdrome until more permanent accommodation could be found.

After the evacuation of Dunkirk beaches using many Southampton-based craft, over 2,000 dejected French troops arrived in port. Some did not even have boots, let alone weapons. The crowded Guildhall now became a distribution centre, and the dock workers and people of Southampton responded warmly to the urgent appeal for help with spare food and clothing. Sadly, the PS *Gracie Fields* that had been sent from Southampton to Dunkirk to assist with the evacuation did not return. The news of this loss of a favourite local paddle steamer was greeted with great shock around the Port of Southampton community.

As the Germans entered the French port of Cherbourg, a number of Scott Paine-built French Navy motor torpedo boats (MTBs) just managed to escape and cross the Channel to Southampton. The exhausted crews with their MTBs arrived at Hythe, and after a few days they were enlisted into British coastal defence flotillas.

The author's father and grandfather were present in the docks during the first German Luftwaffe air raid on 20 June. The author's father, John Britton, gazed up to the sky and pointed out to the author's grandfather what he thought was a strange set of birds. Almost simultaneously, the air raid sirens sounded and the shelters quickly filled up, with the doors firmly locked. Seeking shelter, Alfred and John Britton foolishly decided to take cover under a horse-drawn milk float. As soon as the German aircraft appeared over the docks the anti-aircraft batteries began to open fire and the first bombs began to fall. Taking fright, the horse tethered to the milk float took to his hooves and galloped off! This left the two members of the Britton family rather exposed as pieces of shrapnel began to rain down from the heavens. Taking to their heels they found alternative shelter under an empty barrel.

These first air raids not only shocked the local population, but they also caused considerable damage to buildings and facilities, including the international cold storage building at Berth 40. This facility, which was hit on 13 August 1940, contained 2,345 tons of butter which burned for ten days, creating toxic fumes and smoke. When the last embers of the fire were extinguished, the remains of the butter offered a breeding ground for flies which proceeded to plague the docks and shipping.

Over the weekend of 30 November to 2 December 1940, Göring's Luftwaffe target-bombed Southampton Docks resulting in the quayside sheds at 103 and 104 Berths being completely destroyed. The Red Funnel paddle steamers, *Duchess of Cornwall* and *Her Majesty*, received direct hits and were sunk at their moorings. The *Duchess of Cornwall* was refloated, repaired and returned to service, though it was said that she was never the same mechanically, as she subsequently suffered from sporadic failures in service. The port was temporarily closed as it was felt by Whitehall that it was, 'too dangerous for merchant shipping'. Subsequently, merchant ships were temporarily diverted to other ports in the west. The tug *Canute* was also sunk on 28 December, and during a bombing raid on the Thornycroft yard, the destroyers *Norseman* and *Opportune,* which were under construction, were severely damaged. It was recorded that the *Norseman* was almost split in two by a direct hit. In all, during the war there were a total of sixty-nine air raids on Southampton Docks resulting in twenty-three transit sheds and warehouses containing vital supplies being destroyed.

The spirit of those working in Southampton Docks could not be broken, however, and there were many untold acts of heroism. One such act of bravery could not go unnoticed: dock labourer William Wyatt was raised in a crane sling onto the roof of a blazing building to rescue an injured anti-aircraft gunner who was lying amidst exploding shells. Not only was he given a loud cheer by fellow workers when he successfully completed the gallant rescue, but his bravery was recognised with the award of the George Medal.

It is a little-known fact that, during the hours of darkness in the spring of 1942, the King George V Graving Dry Dock was used for training the commandos who were to take part in a secret night-time raid on the French

▲ Locally produced landing craft are assembled in the New Docks in late May 1944 ready for the 6 June invasion of Normandy. (Associated Press)

port of St Nazaire. The KGV Dock was very similar in design and construction to the Normandy St Nazaire Docks and the commandos were able to familiarise themselves with what potentially lay before them. The commandos practised descending the stairs of the pumping chamber in darkness and planting explosive charges against the pump mechanism and gate-winding machinery. This practice came into its own when the obsolete Royal Navy destroyer HMS *Campbeltown* rammed into the Normandy dock gates and exploded while the commandos destroyed the dock machinery.

Later in the war, the KGV Graving Dry Dock was used to help construct the Bombardon floating breakwaters, which were to protect the Mulberry Harbour during the Allied invasion of Normandy.

As the tide of the war began to change in favour of the Allies, planning and preparations were made for the Port of Southampton to play a key part in the invasion of Europe, in what became Operation Overlord. As early as April 1943, the first landing-craft trials secretly took place. By October 1943, the build-up of troops and supplies commenced with the arrival of Force J and the allocation of twenty-four berths to the force. By March 1944, American units began flooding into the port ready for D-Day preparation. The roads leading into Southampton were like one gigantic traffic jam extending back 40 miles inland. Similarly, the railway network was backing up with train after train of troop specials and freight trains containing vehicles and equipment.

Indeed, it is known that the Great Western cross-country railway line from Didcot, Newbury and Winchester was like one gigantic siding with troop trains stacked for miles inland almost buffer to buffer. The locomotive footplate crews were ordered to sleep and eat on their steam engines and remain on duty until delivery of their loads to the docks. The whole area went into lockdown with strict tight security.

The final loading of landing craft in Southampton commenced on 31 May 1944 and was completed on schedule when Prime Minister Winston Churchill visited to inspect progress. On Monday 5 June, the order was given and the vast armada sailed to a rendezvous point beyond the Isle of Wight to link up with other ships from Plymouth and Portsmouth. Between D-Day and the end of the Second World War, Southampton Docks worked flat out to process out to Europe some 3.5 million troops, 770 steam locomotives with 21,000 carriages and wagons, 39 complete ambulance trains, 22 breakdown trains, 16 mobile workshops to help convey supplies on the destroyed French SNCF railways, 257,680 vehicles and 2.5 million tons of stores. Returning the opposite way into Southampton were 185,273 prisoners of war.

Although the initial assault force from Southampton was mostly British, over two million American GIs sailed from the port in the following months. The people of Southampton and the docks made a lasting impression on US Forces. Consequently, the American Medal of Freedom was awarded to Councillor Stranger to reflect the contribution of the city. By the time of the formal declaration of peace on Tuesday 8 May 1945, it is fair to say that the Port of Southampton had given its all to win the war. Although excitement filled the air with celebrations, the docks and those who worked in them were totally exhausted. The structure, railway system and facilities within the docks were in need of a great deal of investment as so much was worn out beyond economic repair.

◀ Members of the US 16th Anti-Aircraft Artillery Group are shown as they disembark at Southampton. (Associated Press)

6

POST-WAR

Peacetime brought a renewed optimism for the future of the Port of Southampton. The landing hards constructed for D-Day were removed and all signs of military presence began to disappear from the local scene. For the discerning eye there was still evidence of the years of conflict as a section of the Mulberry Harbour was in use as Red Funnel number 2 pontoon by the Royal Pier. Furthermore, for those taking a day trip with their car to the Isle of Wight on Red Funnel, a converted tank landing craft was in use. The post-war fulfilment of peace truly arrived at Southampton with the arrival in port of the Cunard liners RMS *Queen Elizabeth* and RMS *Queen Mary*. The *Elizabeth* entered Southampton on her maiden visit on 20 August 1945, five years after completion by John Brown's on the Clyde in Scotland. She sailed to Canada, taking excited troops back from Europe to their homeland, and returned with weary German prisoners of war. On one voyage, she returned home carrying the Magna Carta! In 1946, the *Mary* made six voyages carrying 9,000 GI brides from Southampton to new lives in the United States.

Both of the Cunard Queens required extensive refits after the years of wartime service as troopships, which took them to the Far East in addition to service in the North Atlantic. In June 1946, the RMS *Queen Elizabeth* entered the King George V Graving Dock for major conversion and refit to become a passenger liner. She looked resplendent on leaving for her official peacetime maiden voyage on 16 October with 2,288 passengers. Next it was the turn of the RMS *Queen Mary* to enter the Graving Dock. She had a new stem fitted to repair the damage incurred when the liner accidentally collided with and sunk the cruiser HMS *Curacoa.* Many of her original fixtures, furniture, fittings, carpets and artwork were retrieved from storage in New York and Australia, as well as Lyndhurst and Brockenhurst in the nearby New Forest. To great cheers from enthusiastic crowds of well-wishers, the *Mary* set sail from Southampton on 31 July 1947.

On 1 January 1948, the railways were nationalised and the Southern Railway became the Southern Region of British Railways. The British Transport Commission took over the management of Southampton Docks and one of their first acts was to implement the construction of a new passenger terminus for passenger liners using the Ocean Dock. On 31 July 1950, the new £750,000 Ocean Terminal was opened by the prime minister, the Rt Hon. Clement Atlee MP. This iconic building was Southampton's symbol of emergence from austerity.

This new 1,200ft-long art-deco-style building elegantly matched the grandeur of the prestigious liners that were to use it, with two luxurious passenger lounges for first and cabin class on the upper floor. Visually, the seaward end of the building had a ship's bridge effect. The terminal was provided with electric heating and air conditioning – at the time a revolutionary feature, but which has since become the norm in public buildings. Passenger accommodation in the reception halls consisted of island settees comfortably upholstered in plush Vaumol hide. Amenities in each hall included a refreshment buffet, telephone bay with operators in attendance, a flower shop, bank, book stall, writing room and a novelty ice-water fountain to make the Americans feel at home. For many years the *Echo* newspaper shipping correspondent, Keith Hamilton, worked at the WHSmith book stall. Each morning he would open up and switch on the illuminated globe above the stall and comment: 'I'm lighting up the world.' The BBC also had a special studio room where VIPs passing through the Ocean Terminal could be interviewed by the likes of the former Welsh war correspondent, Wynford Vaughan Thomas. Wynford related to the author that

➤ Seven-year-old Stephen Rawlings (left) and nine-year-old Pauline Farmer (right) play a game of king-sized deck draughts oblivious to the sailing departure of the SS *United States* on 16 June 1959. (Associated Press)

once the stars stepped into the studio for an interview, the door firmly closed and the world waited outside while an exclusive interview was recorded.

Stepping into this palace of the port was an exciting experience, with a buzz of activity. There was always the slim possibility of meeting someone famous: royalty, a Hollywood film star like John Wayne or Elizabeth Taylor, a notable politician or a distinguished statesman like Winston Churchill. Gossip would quickly circulate around the Ocean Terminal if there was someone of note on board an inbound liner, and a flock of reporters and cameramen would descend like waiting vultures. The terminal had a distinctive and unforgettable aroma of Cuban cigars, expensive perfumes and beeswax polish which filled the nostrils. In this chic setting, passengers expected the best and demanded the highest service from busy porters who toiled endlessly, wheeling mountains of luggage and toppling trunks.

A glance across the first-class lounge revealed attractive fashion icons proudly strutting around like glamorous peacocks. Welcoming bouquets of sweet-scented red roses would be presented by adoring admirers. Dramatically, they would be acknowledged with the words, 'Oh darling ...' Smartly uniformed hotel representatives from Claridge's, the Savoy or the Dorchester would be patiently waiting to greet their first-class guests and gently whisk them through customs and immigration control to the waiting Rolls-Royce or Bentley. Meanwhile, departing American passengers would be sipping cocktails while reminiscing about William Shakespeare, Stratford-upon-Avon, Buckingham Palace and their trip to 'little old England' in these sophisticated surroundings. These elegant and stylish liners were like sumptuous theatres and left a lasting impression of luxury and spaciousness.

The new Ocean Terminal boasted electrically operated aluminium telescopic gangways linking the liners and the terminal at first-floor level. The six gangways were arranged in pairs and could travel along a platform in order to positon them in the shell doors of the docking liners. Furthermore, the gangways could be elevated or lowered to correspond with the level of the ship's door. After customs clearance, passengers would descend to a 1,000ft-long island railway platform, which could simultaneously accommodate two full-length Waterloo-bound boat trains. British Railways Eastleigh Locomotive Shed provided their proudly polished crack Lord Nelson and West Country Class, which gently simmered with clouds of white steam at the end of the platform, waiting to haul the 'Ocean Liner Boat Express'. These elite trains were often composed of a mixture of chocolate- and cream-liveried first-class Pullmans and malachite-green liveried Southern Region Bulleid carriage stock.

➤ A wildcat walkout strike of Southampton seamen at Dock Gate Number 8 on 23 June 1955. (Associated Press)

For visitors to the Ocean Terminal, a high-level balcony was provided on the building's airy roof. It extended the whole length of the terminal on the quayside, with a covered enclosure in the centre. This provided an excellent point to say farewell to departing passengers in the adjacent liners and provided some spectacular panoramic views of the docks. In wet and windy weather one had to hold tight on to hats and scarves, as they would often be blown away into the shadowy depths below!

On 3 July 1952, news arrived in Southampton that the new United States Line superliner, the SS *United States*, had sped away from United States Lines' Pier 86 in New York City and was heading out into the ocean to fulfil her destiny as the fastest liner on the North Atlantic. Reports were received that on her maiden eastbound voyage, her master, Captain Harry Manning, commodore of the United States Lines, encountered fog during the first day of the voyage and had cautiously reduced speed. However, as soon as the fog lifted, he ordered full power and the liner responded well. She performed to all expectations in heavy swells and for a time exceeded 36 knots. Crossing the finishing line at Bishop's Rock, England, the SS *United States* had easily achieved the record held for fourteen years by the British Cunard RMS *Queen Mary*, arriving in an unprecedented three days, ten hours and forty minutes. The victorious SS *United States* triumphantly entered Southampton on 7 July 1952 at the end of her eastbound crossing. The author's father recalled that

she displayed signs of her high-speed dash across the Atlantic, as her fresh new paint had been peeled back along the water line where the superliner had brushed against the waves.

On the return westbound from Southampton, Captain Manning pointed the liner's streamlined bow back toward the United States and once again broke all former records, arriving in New York in three days, twelve hours, twelve minutes, with an average speed for the westbound crossing of 34.51 knots. Although there was a congratulatory welcome by the people of Southampton, the speed and achievement of the new racehorse of the Atlantic was met almost with disbelief, especially by Southampton-based Cunard Queen crews.

In 1955, P&O and Orient Lines ordered what were to be their last passenger liners, *Oriana* and *Canberra*. They were designed for the Australia run and these fast ships were to reduce the length of voyages to just three weeks. In readiness for the maiden voyages of the *Oriana* and *Canberra,* the P&O and Orient Lines companies announced in 1960 that they were going to transfer their services from Tilbury to Southampton. The *Oriana* reached a recorded top speed of 30 knots during her trials, and her maiden voyage from Southampton to Sydney was in December 1960. She was briefly the largest P&O and Orient Lines passenger liner in service on the UK to Australia and New Zealand route, until the introduction of the 45,733 ton *Canberra* in 1961. The *Oriana*, nevertheless, remained the fastest ship in the fleet and was the holder of the 'Golden Cockerel Trophy'.

During 1961, P&O bought out the remaining stake in Orient Lines and rebranded its passenger operations as P&O-Orient Lines. The journey halfway around the world to Australia, assisted by the Australian Government, who encouraged UK immigrants on a one-way £10 ticket, became known as the 'Ten Pound Pom' scheme. At £110 less than a tourist ticket, travellers had to stay at least two years in Australia. Once the ship arrived in Sydney many decided never to return, and by the mid-1970s more than one million Brits had taken up the offer to sail down under.

In 1956, the experience of the Ocean Terminal led to the construction of a new two-storey terminal in the Western New Docks for the Union-Castle Line's service to South Africa. It was opened on 25 January 1956 by Mr G.P. Jooste, the High Commissioner for South Africa, with passengers for the *Edinburgh Castle* being the first to use the new facility. Also in 1956, work commenced on the construction of a new cold store for the International Cold Store & Ice Company at 108 Berth. This opened two years later on 16 July 1958 with the arrival of the *Brisbane Star* from New Zealand.

Port facilities continued to be improved in the late 1950s when reconstruction commenced of the pre-war passenger waiting hall between 105 and 106 Berths. The new accommodation was officially opened by Field Marshall, The Rt Hon. Viscount Slim on 29 November 1960, with passengers from Orient Lines' *Oriana* the first to use it on the ship's maiden voyage to Australia on 3 December.

in 1961, British Railways decided to transfer their Channel Islands passenger services from Southampton to Weymouth. The Southampton–Channel Islands' service sadly came to an end with the arrival of the *Isle of Guernsey* on 12 May that year. British Railways added to the bad news three years later; they ceased ferry services to Le Havre from Southampton On 9 May 1964 and to St Malo on 27 September 1964. The closing of these services meant that British Railways ended all passenger services from the port, but freight services continued to operate until 1972.

The vacuum left by the departure of British Railways' cross-Channel services was filled by the introduction of new private continental operators. The Inner Dock was filled in, the Outer Dock redeveloped, and two new drive-on, drive-off car ramps were built. New buildings for a reception hall were constructed and large areas were bulldozed and cleared for car and vehicle parking. The completed car-ferry facilities were opened on 3 July 1967 by HRH Princess Alexandra, and named after her.

In January 1963, management of the Port of Southampton was transferred from the British Transport Commission to the British Transport Docks Board. With the perceived decline in passenger shipping figures owing to the impact of the jet aeroplane, the new Docks Board determined that the future of the port lay in cargo operation. Consequently, it was decided to plan and construct a new container dock between the King George V Graving Dry Dock and Redbridge Causeway. The government-sponsored container facility was planned to provide thirty deep-water berths and necessitated considerable land drainage and reclamation at a cost of £60 million. The first new container quay opened on 28 October 1968 and the inaugural ship to arrive was Dart Line's *Teniers*.

In the 1950s and 1960s, Southampton developed a reputation for strikes and militancy, extracting rises in pay or cuts in hours. Dock strikes and seamen's strikes had much the same effect on operators, both of them stopping shipping movements. Localised Southampton strikes caused ships and liners to temporarily transfer services to Plymouth. Others cancelled docking in the UK and terminated scheduled transatlantic services at Le Havre or Cherbourg. This encouraged many first-class passengers to opt to travel by air.

⌃ In this wonderful panoramic aerial view taken in the early 1960s, looking north towards the River Test estuary showing the entire docks area, we can identify the twin-funnelled Cunard *Mauretania* berthed at Berths 46 and 47 in Ocean Dock. Alongside her is an Alexander Towing Company tug at 45 Berth and an unidentified British Railways cross-Channel ferry. Also worthy of note are a line of P&O and Union-Castle Line ships in the New Docks. The scene is complemented by the Marchwood Power Station chimneys at the top left of the picture. (Esso/Britton Collection)

⌃ In 1954, photographer Pursey Short flew over Southampton Docks in a De Havilland Rapide aircraft and took a series of remarkable aerial views of the construction of Marchwood Power Station and the New Docks. The twin iconic chimneys were to become a dominant feature on the local skyline until the 1980s. (Pursey Short/Britton Collection)

◄ The P&O *Canberra* rests in the King George V Graving Dock while undergoing a refit. The enormity of the dry dock can be gauged by the diminutive figure pictured at work in the foreground on the floor of the dry dock. (Gwilym Davies/ Britton Collection)

❯ At the extreme west end of the New Docks was the King George V Graving Dock, which was the largest of Southampton's six dry docks. Completed in July 1933, this dry dock was designed to accommodate liners of up to 100,000 tons. The site of the dry dock was formerly tidal mudland, part of the bay of 400 acres reclaimed to form the New Docks. 'The concrete floor of the dock is 25ft thick and in the floor and walls 456,000cu. yd of concrete were used,' Uncle Fern Bussell, former KGV Dry Dock painter, would proudly reveal to the author's family. In John Cox's picture we see the P&O liner *Oriana* receiving a hull inspection, scrape and paint in May 1963. (John Cox)

⌃ Home Lines *Queen Frederica* is pictured entering the King George V Graving Dock for a full refit. This wonderful ship was launched on 26 June 1926 as the SS *Malolo* for the Matson Navigation Company. The 17,226-ton cruise ship was later sold to Chandris Lines and was eventually scrapped in July 1977. (Gwilym Davies/Britton Collection)

⌃ As water is slowly pumped out of the King George V Graving Dock, a team of painters row into position to commence barnacle-scraping the hull of the Home Lines *Queen Frederica*. (Gwilym Davies/Britton Collection)

⌃ A dramatic photograph of the Shaw Savill Line *Northern Star* in dry dock as the swirling water is pumped out ready for the liner's refit. The 24,700-ton *Northern Star* was only thirteen years old when she was withdrawn to be broken up. She was the victim of massive changes which also greatly reduced most other passenger shipping fleets, namely the sharp rise in fuel and operating costs, plus the increased popularity of cheap air travel. During her brief career, she was used on Shaw Savill's 'round the world' service from Southampton. With an overall length of 650ft and a width of 82ft, *Northern Star* followed the same revolutionary design as her sister ship, *Southern Cross*, with the engines aft and no cargo holds. (Gwilym Davies/Britton Collection)

▲ Cunard RMS *Queen Mary* has her three funnels caged in scaffolding while undergoing refit in the King George V Graving Dock. The gigantic quayside crane is pictured removing each of her lifeboats for refurbishment, repainting and safety certification following inspection. Work on the refits of vessels in the dry docks would continue around the clock. Note the classic cars and motorcycle and sidecar parked in front of the massive Cunard Queen liner. (Britton Collection)

▲ The P&O Line's *Oriana* dwarfs the painters who are busy at work on their precarious ladders at the base of the bow stem. Launched on 3 November 1959, she was the last of the Orient Steam Navigation Company's ocean liners. When the *Oriana* first appeared at Southampton she looked resplendent in her owners' traditional corn-coloured hull. When the Orient Line was absorbed into the P&O group in 1966, the Oriana was refitted and recarpeted in Southampton. The author visited his neighbour, Sid Deeming, who worked on Oriana, and was very impressed with the new décor. (Britton Collection)

▲ The flagship of the Cunard Line, the 83,673-ton RMS *Queen Elizabeth*, looks majestic as she undergoes refit in the King George V Graving Dock. (Britton Collection)

▲ The commemoration stone to mark the opening of the King George V Graving Dock by His Majesty King George V with Her Majesty Queen Mary on 26 July 1933. (Britton Collection)

▼ The unrivalled and serene beauty of the SS *United States* is seen in all her glory as she is gently turned by a brace of Alexandra Towing Company tugs after sailing from 107 Berth at Southampton in March 1958. (Gwilym Davies/Britton Collection)

▲ There were three complete sets of linen aboard the SS *United States*. Additionally, one set of linen was at the laundry in New York and one at the laundry in Southampton. Soon after arrival at Southampton, 'the dirties' were unloaded ashore. The average number of bags of soiled linen landed per voyage was 650, both at New York and Southampton. In early 1969, the cousin of the author, Anthony Webb, was summoned urgently onto the SS *United States* immediately after docking at Southampton to repair a broken drum in the onboard laundry. He recalls working continuously until the job was completed minutes before sailing. Commodore Alexanderson came down personally to thank Anthony for rectifying the problem. (Gwilym Davies/Britton Collection)

⌃ Captain Jack Holt, the Trinity House pilot on board the SS *United States*, relays final commands to all hands and the captains on board the six assisting tugs, 'Make ready to sail' on 26 August 1954. (Pursey Short/Britton Collection)

⌃ Ordered chaos! The six assisting Red Funnel and Alexandra Towing Company tugs appear to be dancing around to new positions in order to escort the SS *United States*. (Pursey Short/Britton Collection)

‹ A sea-level view of the departure of the SS *United States* in March 1958 with the Alexandra Towing Company tug *Sloyne* leading the way. This tug was built in 1928 and weighed a mere 300 tons, but had 900hp. (Gwilym Davies/Britton Collection)

➤ Simply spectacular. Crowds watch the departure of the SS *United States* from Mayflower Park on 26 August 1954. Behind is the Southampton Civic Centre, a landmark on the skyline. (Pursey Short/Britton Collection)

➤ Taken on 26 August 1954, this impressive aerial view of the SS *United States* sailing from Southampton reveals that she is being followed by four boatloads of admirers, snapping action shots from their camera lenses. A beautiful reflection from the world's fastest ocean liner can be seen beside her. (Pursey Short/Britton Collection)

◄ The inbound SS *United States* passes the flagship of the Cunard Line, the 83,673-ton RMS *Queen Elizabeth*, outbound for Cherbourg and New York. This rare action shot was taken from a Blue Funnel pleasure boat off Hythe Pier in 1968. The fastest passenger ocean liner in the world is passing the largest passenger ocean liner in the world. (Barry Eagles)

▲ The *Nieuw Amsterdam* was nicknamed 'The Darling of the Dutch'. Launched in April 1937, she had a very interesting career: After the German invasion of Holland during the Second World War this liner was put at the disposal of the British Ministry of Transport by the exiled Dutch Government. The *Nieuw Amsterdam* is pictured in the New Docks on a transatlantic voyage from Rotterdam to New York. In 1957, she was fitted with air conditioning and her hull was painted grey. (David Peters)

◄ Maintenance of ocean liners was never-ending, and here we see the *Nieuw Amsterdam* receiving a quick lick of paint while tied up in the New Docks. (Britton Collection)

▲ Top: Refuelling vessel pictured at 106 Berth in the New Docks, in August 1959, is the German-registered *Haseldorf*. The Port of Southampton is perceived as being a passenger terminal, but in fact it was and is to this day a very busy commercial port; Bottom: The heavily laden refuelling vessel *Esso Lyndhurst*, which is low down in the water, is seen making her way slowly up Southampton Water past the New Docks. She is to replenish a P&O Line cruise ship with Bunker C oil. The backdrop is Marchwood Power Station. (Both Britton Collection)

◄ The P&O Line *Orcades'* funnel was similar in appearance to that of her sister ship *Oronsay*. The first-class accommodation on board was almost identical too. This 28,396-tonner was built in 1948 by Vickers Armstrong at Barrow. When refitted at Belfast in 1959, *Orcades* was air-conditioned and equipped with a new tavern and swimming pool for the first-class passengers. This would have been a very welcome addition when cruising to Australia. (Gwilym Davies/Britton Collection)

▲ The photographer Gwilym Davies has captured the plain black funnel of *Woodlark* in August 1959, docked at 104 Berth. Note the distinctive badge on the funnel of the General Steam Navigation Company (part of the P&O Group). *Woodlark* regularly sailed to Danish ports, as well as Gothenburg and Rotterdam. (Gwilym Davies/Britton Collection)

▲ It was often said that the single funnel of the *Caronia* was the largest on any British liner. This 715ft cruise liner had a maximum speed of 22 knots and became known in Southampton as 'the millionaire's yacht'. Consequently, tips to the crew were much better on the *Caronia*, and it followed that there was no shortage of volunteers to staff this magnificent liner. The author can confirm that a visit on board was truly like stepping into paradise – pure luxury. One can close one's eyes and recall the *Caronia* had a most distinctive smell of polish, Brasso and new carpets! (Arthur Oakman/Britton Collection)

▲ Bibby Line's and Britain's second troopship was the *Oxfordshire*. Built in 1957 by Fairfield of Glasgow with a mean speed of 21 knots, she was a familiar sight in Southampton Docks and is seen here at 101 Berth. (Graham Cocks/Britton Collection)

➤ Photographer Austen Harrison took this view when taking a 'tour of the ship' for a last look at the Cunard RMS *Queen Mary* at 107 Berth on 10 October 1967, prior to her final voyage to Long Beach, USA. At this stage she was being cleared of items to be retained by Cunard and stocked up with provisions for the epic voyage ahead. (Austen Harrison/World Ship Society)

⌄ An impressive line-up of ships at 101 Berth on Remembrance Day in 1972: *Clan Ramsey, Clan Robertson* and *Rothersay Castle*. However, this magnificent sight only tells half the story, for they are all laid up awaiting disposal. (John Wiltshire)

◄ With the end of the Seamen's Strike in sight on Friday 1 July 1966, the twin chimneys of Marchwood Power Station gently smoke across the Test estuary. Pictured from Mayflower Park in 101 Berth are the Union-Castle Line ships (left to right): *Good Hope Castle, Reina del Mar* and *Edinburgh Castle.* (Britton Collection)

◄ With the end of the Seamen's Strike in sight on Friday 1 July 1966, pictured from Mayflower Park in 101 Berth are the Union-Castle Line ships (left to right): *Good Hope Castle, Reina del Mar* and *Edinburgh Castle.* (Britton Collection)

⌃ An impressive rooftop view of the funnels of the for-sale Cunard RMS *Carmania* and *Franconia*. (Gwilym Davies/Britton Collection)

⌃ For sale: Ex-Cunard liners RMS *Carmania* and RMS *Franconia*. Offers over £1 million each. Viewing at 101 Berth by appointment with H.E. Moss & Company Ltd. (John Goss)

⌃ South Africa-bound *Pretoria Castle* of the Union-Castle Line waits patiently like a solitary soldier guarding 101 Berth. (Pursey Short/Britton Collection)

◄ Union-Castle *Windsor Castle* was the largest ship ever built by the Union-Castle Line. Built by Cammell Laird of Birkenhead, she was launched by Her Majesty Queen Elizabeth, The Queen Mother on 23 June 1959. *Windsor Castle's* maiden voyage was on 18 August 1960 from Southampton to Capetown, and the author has memories of the lively music played by a military band, the cheering crowds and the streamers at the dockside as he watched it sail away. (A.E. Bennett)

◄ This slide shows the heyday of Southampton liners in the early 1960s. On parade in this impressive New Docks line-up is the Union-Castle Line *Pendennis Castle, S.A. Vaal* (former Union-Castle Line *Transvaal Castle*), British India Line SS *Nevasa*, P&O Lines *Oriana* and *Chusan*. (Pursey Short/Britton Collection)

▲ An early telephoto-lens view from Hythe of the Union-Castle Line *Pendennis Castle*. Note the train of fitted, covered wagons on the quayside. (Gwilym Davies/Britton Collection)

▲ While preparing to depart from 101 Berth, we have a view from the bridge of Union-Castle Line *Windsor Castle* looking across Mayflower Park. The twin funnels of RMS *Queen Elizabeth* protrude from Ocean Dock. (G.R. Keat/Britton Collection)

▲ Here is a classic scene with horse-drawn street cleaning, looking across Mayflower Park with Union-Castle Line *Windsor Castle* in 101 Berth on the right and the twin chimneys of Marchwood Power Station dominating the skyline. (G.R. Keat/Britton Collection)

▲ This typical summertime view, taken in June 1966, shows tourists and locals relaxing on the grass watching the endless procession of shipping at Mayflower Park. The scene shows the Union-Castle liners *Reina del Mar* and *Edinburgh Castle* tied up at 101 Berth. However, also worthy of note are the long line of classic cars. On the right are a Hillman Minx, Vauxhall Viva, Ford Zephyr, Commer van, MG, Austin Mini and many more. (Norman Roberts/Britton Collection)

▲ This sequence of pictures shows the departure of the Union-Castle Line *Stirling Castle*. This was a weekly service to Cape Town leaving Southampton every Thursday at 4 p.m. and arriving in Cape Town at 6 p.m. on the fourteenth day of the voyage. The return voyage timetable was to sail from Cape Town every Friday at 4 p.m. and berth at Southampton a fortnight later at 6 a.m. The time taken on the passage was thus thirteen days and fourteen hours. We see *Stirling Castle* moored at 101/102 Berth ready to depart. Next, the harbour master's launch *Apace*, with Mr S.N. Finnis, dock manager, on board, arrives to supervise proceedings. No fewer than five tugs assist from the locally based Alexander Towing and Red Funnel fleets. (Pursey Short/Britton Collection)

▲ The 1951-built Elder Dempster Lines *Aureol* is seen raising steam for a smoky departure sailing from Southampton New Docks. She was launched on 28 March 1951 and commenced her maiden voyage on 3 November 1951. Her curved bow, terraced superstructure, tripod mast and cruiser spoon stern made her one of the handsomest ships to call at Southampton. (John Wiltshire)

Shaw Savill Line

SOUTHAMPTON

➤ Almost lost in the autumnal morning mist with the twin chimneys of Marchwood Power Station across the opposite side of the water at 101 Berth is the Chandris Line *Ellinis*. This vessel had an interesting history and was built as the *Lurline* for the Matson Navigation Company. During the war she was converted to a US Navy troopship. At the time of this picture *Ellinis* was used for cruising from Southampton to the Mediterranean. (Britton Collection)

➤ The P&O Line *Arcadia* was the largest P&O liner ever to be built on the Clyde. Launched on 14 May 1953, she is reputed to have cost £6.5 million. She was a beautifully proportioned liner and soon gained a fine reputation for efficiency and comfort. Gwilym Davies has captured her through his lenses in September 1958 flanked by cranes standing like sentinels. (Gwilym Davies/Britton Collection)

This sequence of remarkable pictures captures the departure of *Arcadia* from Southampton bound for Sydney on 21 August 1954. We see her assisted by escorting tugs: the straw-colour-funnelled Alexander Towing Company tug tender *Romsey*, tug *Brambles* and Red Funnel tug tender *Paladin*. (Pursey Short/Britton Collection)

▲ The 29,614-ton P&O *Iberia* rests patiently at 101 Berth. Beyond are two unidentified Union-Castle and P&O liners. (Graham Cocks/Britton Collection)

➤ Funnel focus, the unique and distinctive *Himalaya* on 31 August 1961. (Gwilym Davies/Britton Collection)

➤ When built in June 1949, *Chusan* was the largest P&O liner for the Far East service. A visit aboard revealed she had two swimming pools for her 1,005 passengers and was air-conditioned. At the time the picture was taken, *Chusan* was bound for Port Said, Aden, Bombay, Melbourne and Sydney. The author recalls the ship's children's room was one of the best of any liner and was the first on any Southampton-based liner to have the children's toy Lego. It was a difficult task for the author's father to drag his son away! (John Wiltshire)

⋏ Left: One can only appreciate the sheer size of a liner by gazing up at the front of the sharp rake stern. Here we can gaze in awe and wonder at the coat of arms on the stem of P&O *Oriana* on 25 March 1961. (Gwilym Davies/Britton Collection); Right: The Alexandra Towing Company tug *Ventnor* heaves *Oriana* away from her moorings in August 1979. (Norman Roberts/Britton Collection)

◅ A golden glint on the P&O Line *Oriana*. (David Peters)

Without doubt one of the most popular liners to frequent Southampton was the P&O *Canberra*. She was affectionately known as the 'Great White Whale', due to her famous exploits in the Falklands War. When she first entered Southampton, many, including the author, felt she was unconventional in design. However, with time she appealed to everyone who had contact with her. Photographer Pursey Short took this series of photographs of her as a tribute to one of his favourite liners. (Pursey Short/Britton Collection)

▲ An electric elevating platform truck is seen bringing in a hoist from the quay containing Cape apples from South Africa. (British Railways Board)

▲ South African wool was landed from the weekly Union-Castle liner on the upper floor of 102 Berth shed. It is seen here awaiting distribution to the wool mills of Yorkshire and transhipment to the Continent. (British Railways Board)

◄ Sunrise arrival at the New Docks on 1 September 1951 as the veteran Cunard RMS *Franconia* prepares to dock escorted by an Alexandra Towing Company tug and a Red Funnel tug. (Pursey Short/Britton Collection)

7

1966 SEAMEN'S STRIKE

The Port of Southampton was paralysed in 1966 by the National Union of Seamen's first strike since 1911. The strike started on 16 May with the aim of securing higher wages and reducing the working week from 56 to 40 hours. It ended on Friday 1 July after causing great losses to the shipping companies operating to and from the port. The political importance of the strike was enormous as it disrupted trade and had an adverse effect on the country's balance of payments, threatening to undermine the British Government's attempts to keep wage increases below 3.5 per cent. The strike was widely supported not only in Southampton, but in other British ports at Liverpool and London.

The 1966 Seamen's Strike caused an almighty seize-up of British shipping, so that when a British-crewed vessel arrived in the Port of Southampton, it was immediately tied up at berth and strikebound. For the shipping enthusiast it was a once in a lifetime chance to witness a feast of ocean vessels, as the great British fleets of liners remained tied up and idle. Cunard watched helplessly as their giant Queens – *Caronia, Franconia* and *Carmania* – were tied up and began to lose huge sums of money for the company. Similarly the Union-Castle Safmarine saw all their passenger ships on the South African 'Cape Run' come to a standstill in the port. One by one, ships joined the gathering as they entered port and tied up, some doubling up at berth and even treble-banked with three Union-Castle liners at one berth! Two of the P&O fleet, the *Canberra* and *Arcadia,* along with Royal Mail Line's *Andes* also found themselves strikebound. Some berths were deliberately kept unoccupied for the use of foreign flag vessels calling at the port.

They say 'every cloud has a silver lining' and it is 'an ill wind that blows nobody any good', and so it was with the Seamen's Strike of 1966. Local sightseeing pleasure-boat operators made a fortune during the weeks of the strike by taking vast crowds of visitors around the docks for a review of the British Merchant Navy Fleet (the author being one such visitor). At the end of the strike on Friday 1 July the veteran maritime enthusiast Anthony E. Bennett from Norwich made a leisurely visit of the docks recording every vessel docked:

SHIPPING DOCKED IN THE PORT DURING 1966 SEAMEN'S STRIKE

OLD DOCKS

Berth	Ship	Owner
7	*Viking I*	Thoresen Car Ferries
7	*Viking II*	Thoresen Car Ferries
7	*Viking III*	Thoresen Car Ferries
20	*Northolm*	Oldendorff
22	*Lune Fisher*	Fisher
23	*Elk*	British Railways
25	*Leon*	Empressa Hondurena (foreign-registered en route from Port Antonio)
26	*Rio San Juan*	Elma (foreign-registered en route from Santos)
29	*Twyford*	Risdon Beazley

29	*Moose*	British Railways
30/31	*Southern Cross*	Shaw Savill
30/31	*Franconia*	Cunard
32/33	*Capetown Castle*	Union-Castle
32/33	*S.A. Orange*	Safmarine
34/35	*Clan Robertson*	Clan
36	*Good Hope Castle*	Union-Castle
37	*Chicanoa*	Elder & Fyffes
38/39	*Caronia*	Cunard
40	*Andes*	Royal Mail
41	*Lions Gate*	Johnson
43	*Roxburgh Castle*	Union-Castle
43	*Rowallan Castle*	Union-Castle
43/44	*Queen Mary*	Cunard
46	*Carmania*	Cunard
46	*Port Lyttleton*	Port Line
47	*Reina del Mar*	Union-Castle
49	*Bess*	Wallenius
50	HM ships: *Venturer, Thames and Warsash*	
50	*Camito*	Elder & Fyffes
	Golfito	Elder & Fyffes

NEW DOCKS

Berth	Ship	Owner
101	*Edinburgh Castle*	Union-Castle
	Reina del Mar	Union-Castle
	Good Hope Castle	Union-Castle
102	*Canberra*	P&O
102/103	*Alexandra K*	Merabello SA
104	*S.A. Vaal*	Safmarine
105	*Willemstad*	KNSM (foreign-registered en route Amsterdam–Madeira)
106	*Queen Elizabeth*	Cunard
107	*Rotterdam*	Holland America (foreign-registered en route New York–Rotterdam)
107	*France*	CGT French Line (foreign-registered en route Le Havre–New York)
107/108	*Arcadia*	P&O
110	*Bay Fisher*	Fisher
110	*Dorset Coast*	Coast Lines

DRY DOCKS

Berth	Ship	Owner
5	*Southampton Castle*	Union-Castle
6	*Pendennis Castle*	Union-Castle
7	*Windsor Castle*	Union-Castle

◄ A memory of the Seaman's Strike in June 1967, taken from an inbound Red Funnel ferry from Cowes. The Isle of Wight photographer, David Peters, was on a photographic safari to record the final workings of British Railways Southern Region steam and could not resist recording this scene at Ocean Dock showing the Cunard RMS *Queen Mary*, Union-Castle Line's *Roxburgh Castle* and *Rowallan Castle*, and a whole dock full of strikebound ships. (David Peters)

8

SOUTHAMPTON: THE PLACE TO BE SEEN

In its heyday, many Hollywood film stars, royalty, politicians and celebrities would pass through Southampton to travel on the ocean liners. Many had their favourite liners on which they preferred to travel. Before the Second World War, there was only one way to travel across the Atlantic from Southampton and that was via the Cunard RMS *Queen Mary*. This tradition continued after the war when she was joined by her larger sister, the RMS *Queen Elizabeth*. The Cunard service of the pre-war period from Southampton didn't diminish with the passage of time. Those with affluence and influence, together with the glamorous figures of the screen, queued for a ticket on the Queens. They paid for luxury surroundings and first-class service that started at the Port of Southampton. At the front of the queue was wartime Prime Minister Winston Churchill, who reserved a suite for a well-deserved rest on board the *Queen Elizabeth* in 1945. He arrived at the quayside accompanied by a fleet of vans carrying his crates of vintage wines, artist oil-painting canvases and brushes, cigars, and trunks with a variety of suits and hats.

The Duke and Duchess of Windsor were at first regular travellers aboard the Cunard liners from Southampton. However, from 1952, the Windsors transferred their allegiance to the United States Lines following a dispute with the duchess when she expected Cunard to subsidise her transatlantic ticket.

Four Presidents of the United States passed through Southampton arriving on board the SS *United States*: Harry Truman, Dwight Eisenhower, John F. Kennedy and Bill Clinton. The youthful Bill Clinton, complete with his saxophone, was fresh out of Georgetown and on his way to study at Oxford.

➤ Former US President Dwight D. Eisenhower waves to the crowds from the sundeck of the Cunard RMS *Queen Elizabeth* at Ocean Dock on 18 July 1962. He was returning home to New York with his wife and family. (Associated Press)

On nearly every voyage into Southampton, the SS *United States* carried a distinguished list of celebrity passengers, including such famous Hollywood stars as: Marlon Brando, Gary Cooper, Walt Disney, Bob Hope, Bing Crosby, Judy Garland, Cary Grant, Charlton Heston, Marilyn Monroe, John Wayne and Elizabeth Taylor, to name but a few. The celebrities would always be met by a crowd of cameramen all eager to snap a picture for their paper, and reporters would wait for a memorable quote or word. Comedian Bob Hope, on arriving in Southampton on 18 August 1952 wearing a 10-gallon Stetson hat quipped, 'The *United States* is such a marvellous ship, she could even be

▲ A very sad occasion as Her Majesty Queen Elizabeth The Queen Mother visits the liner she launched, the RMS *Queen Elizabeth*, for a final farewell at Southampton in 1968. (Associated Press)

converted to a passenger ship in a few days! I understand the Cunard Line gave it a twenty-one torpedo salute when it came in to Southampton today. Don't worry, Granddad (Bing) Crosby will be arriving here soon on a slow Cunard liner.'

Straight from Cuckoo Lane, Laurel and Hardy stepped ashore at Southampton in 1947 on their way to reopen the Romney, Hythe & Dymchurch Railway in Kent. During their comical arrival, the press interviews were interrupted by Stan Laurel who vigorously complained that Oliver Hardy was standing on his foot. Some Hollywood stars revelled in the full glare of publicity, while others like Grace Kelly preferred to secretly sneak ashore unnoticed, hidden by a scarf and dark sunglasses.

Some celebrities, like the Christian evangelist Billy Graham, who was on the SS *United States* bound for the Great London Crusade, arrived at Southampton amidst controversy. When he disembarked at Southampton he was mobbed by hostile tabloid reporters. A Labour Member of Parliament had accused Dr Graham of interfering with British politics under the guise of religion. Billy Graham said that he believed God was going to pour out revival on England. As he passed through customs, a US Lines agent and a Southampton dock worker thanked him for coming, as did a Southampton black-cab driver. Billy's spirits were lifted.

There was no shortage of willing stevedores in Southampton to meet the stars off the liners. A smile and a nod were often rewarded with a handsome $20 tip!

On rare occasions, some VIPs and celebrities visited the Port of Southampton for special reasons, other than to travel on a liner. One such occasion was on 6 November 1968 when Her Majesty Queen Elizabeth The Queen Mother paid a final visit to the Cunard RMS *Queen Elizabeth* at 107 Berth. Commodore Geoffrey Marr had given the order to clean and polish his ship for a 'special visitor' who was coming down to say farewell to her ship. When the Queen Mother arrived, the members of the crew drew up a guard of honour, and with the Chairman of Cunard, Commodore Marr and Staff Captain Bill Law, she toured the liner she had launched and named for a final time. When Her Majesty stepped down onto the quayside, she was genuinely saddened to say farewell to her ship.

9

POST-WAR FLYING BOAT SERVICES

On a sad afternoon in December 1958, a chapter in the Port of Southampton's history ended when the last regular scheduled flying boat took off. The era of Aquila Airways in Southampton was at an end. The unique experience of taking off from the Southampton Water by flying boat and climbing up into the clouds over the Isle of Wight, while enjoying superb meals, crisp linen tablecloths and premier service by attractive uniform stewards became a memory. Aquila Airways, which operated from Berth 50 in the Eastern Docks, was started by Wing Commander Barry Aikman in 1948 as BOAC (British Overseas Airway Corporation) began phasing out their flying boat services.

The favourite take-off spot for Aquila flying boats was off Netley, but on occasions, owing to tidal conditions, this was transferred to an area near Calshot. The twice-weekly flying-boat service to Madeira was dubbed the 'armchair route', so comfortable and relaxing were the onboard facilities. Other services were added to Corfu, Italy, Majorca and Montreux. Special charter flights were also a feature, with the longest taking place in 1952 to the Falkland Islands.

Commercial flying-boat services had commenced in August 1919 from Woolston, but really came to prominence in 1936 when the powerful four-engine C Class operated to North America and the Empire routes. During the Second World War the flying boats transferred along the coast to the Dorset base of Poole. In 1948 they returned home to Southampton where BOAC opened a new terminal near the Town Quay. Using the Short Solent and Sandringhams and converted former Coastal Command Sunderland flying boats, they flew nine services a week from Southampton to Australia, South Africa and Japan. BOAC then began replacing the flying-boat routes with land-based flights from London using Lockheed Constellations and Canadair C4 Argonauts. By November 1950, the last scheduled BOAC flight to South Africa was withdrawn.

➤ A Princess flying boat on 13 September 1953 flying over Calshot and Southampton Water. This is one of the very few occasions that this massive flying boat flew. (Pursey Short/Britton Collection)

Great things had been planned to develop the potential of the flying-boat operations from Southampton. In 1945, Saunders-Roe of Cowes on the Isle of Wight were asked to design and build for BOAC a flying boat for use on the transatlantic service. Saunders-Roe received an order for three Princess flying boats which were powered by ten Bristol Proteus turboprop engines, powering six propellers. A rounded bulbous double-bubble pressurised fuselage contained two passenger decks with room for 105 passengers. When BOAC re-evaluated their future plans, they decided to suspend production of the Princess flying boats. It was announced that the RAF would continue their construction as transport aircraft. Sadly only one of the three prototype Princess flying boats ever flew, *G-ALUN*, under the command of test pilot Geoffrey Tyson in 1952. After a few test flights in the Solent and to Farnborough, the project was cancelled and the airframes were put into store pending resale. After lingering for many years through the 1960s, they were unfortunately scrapped as it was discovered that they were badly corroded.

FAWLEY OIL REFINERY

As long ago as 1921 there had been a small oil and petroleum refinery at Fawley in Southampton Water when the Atlantic, Gulf & West Indies Oil Company acquired 270 acres of virgin heathland and mudflats. This site was chosen because a large amount of land was available for development as the area was sparsely populated and because of Fawley's geographical location in the Solent and Southampton Water. In 1951 the ownership of the refinery passed to Esso, which rebuilt and extended the facilities. The planning vision for Fawley made it possible for crude oil to be brought to the site by sea. The proximity of the refinery to Southampton was a crucial factor, as at the outset much of the plant's output was used to supply shipping using Southampton Docks.

At the beginning of the Second World War Fawley was producing 6.7 per cent of the national demand for oil and petrol: 12,000 barrels per day.

However, when hostilities commenced, refining was suspended and Fawley was used as a storage depot. In 1951, after a further 1,200 acres of land were acquired by Esso, the first stage of the expansion plan came on stream. This consisted of primary distillation units, a catalytic cracker and numerous treating units. A chemical plant was completed in 1958. With the advent of new supertankers, Fawley increased capacity to 270,000 barrels per day. Pumping and piping facilities were installed to further increase unloading capacity on the opposite side of Southampton Water. In the 1950s and 1960s regular oil and petrol trains left Fawley to travel along the single line branch bound for the Midlands via the Winchester, Didcot and Newbury line. By the late 1960s much of this traffic had transferred to the road or was distributed by pipeline.

◄ Taken from a passing Cowes-bound Red Funnel paddle steamer, a Red Funnel tug is pictured returning to Southampton past the gigantic Fawley Oil Refinery in April 1961. (Gwilym Davies/Britton Collection)

POST-WAR RAILWAYS AND BOAT TRAINS

During the war the Southern Railway had struggled to use aging former London & South Western Class B4 shunting locomotives. Soaring costs and delays in obtaining replacement boilers for the antique B4 locomotive fleet at Southampton Docks forced the Southern Railway to consider radical ideas for suitable replacements. With the end of hostilities there were many surplus American War Department six-coupled tank locomotives for sale. After testing steam locomotive WD Number 4326, the Southern Railway decided to purchase it, along with thirteen more (plus one more for spares) for £2,500 each. These rather ugly-looking engines had a traditional American appearance with their stovepipe chimneys, sandboxes mounted on the boiler and wide cabs. Their short wheelbase made them ideal for working over the sharp curves around the docks' railway network. These American 'Yankee' tanks acquitted themselves well, but they did have two major failings: the tendency to suffer hot axle boxes and excessive cab heat caused by their large fireboxes. A novel innovation added to the USA tanks was the radio telephone 'cab to shore' system. This controlled the movements around the docks and was unique to Britain. The USA tanks were fitted with whip aerials, radio telephones and turbo generators to provide the required power. Drivers at first objected to this system and were very suspicious, but were mollified when they realised they had an efficient form of communication back to base. They were suddenly able to call up for relief crews, report a fault or remind the control that they were nearing the end of their shift. Right up until the end of steam traction in Southampton in July 1967, 'Yankee' tanks were a familiar sight in the docks and gave excellent service. Examples are preserved at the Keighley & Worth Valley Railway and Kent & East Sussex Railway.

Since the reign of the London & South Western Railway there had always be a tradition of running boat trains from Waterloo to Southampton Docks. In 1952, British Railways Board Southern Region began planning and devising names for the boat train services to Southampton. Locomotive headboards, some adorned with appropriate insignia, were made and sent to Nine Elms shed in London. Similarly, carriage destination boards were made and dispatched to Clapham Junction carriage sidings, ready for use from 1953. The boat trains were named: *The Cunarder* (Cunard Line), *Holland American* (Holland America Line), *South American* (Royal Mail Lines), *Statesman* (United States Line) and *Union-Castle Express* (Union-Castle Line). Additional named boat trains for the two British Railways cross-Channel services were titled *Normandy Express* and *Brittany Express*. From 1957 it was decided to name boat trains connected with Union-Castle Line 'The Springbok', but there is undisputed photographic evidence that the Union-Castle Express headboards were carried by steam locomotives right up to July 1967.

▲ This is the view at Southampton Central railway station down platform with West Country Class steam locomotive 34045 *Ottery St Mary* with an express train for Bournemouth and Weymouth. The London Nine Elms shed engine crew have just handed over the controls of the Bulleid-designed steam locomotive to a Bournemouth crew and seem oblivious to the smoking funnels of the SS *United States*, which is about to sail. (Britton Collection)

◄ Network Rail map of the Southampton Dock area. (British Railways Board Archives)

12

CRANES

One of the secrets of the smooth and efficient working of Southampton Dock was the humble cranes, upon which the speed of loading and unloading of ships largely depended. They were the unsung heroes and workhorses of the port, taken for granted but a vital tool operated by skilled men. From a distance the quayside of the docks could be traced by the lines of cranes, their leaning jibs silhouetted against the sky. As a boy, the author looked on them like a regiment of guards, but to see them in action, close up, gyrating and plumbing the ships' holds, was a memorable sight never to be forgotten.

The author will always remember when the Cunard Queen liners sailed from the Port of Southampton for their farewell voyages. One by one, the cranes around the port bowed their jibs in respect as the great ships passed. Today the traditional dockside cranes as a feature of the port have all but disappeared from the scene.

The first electric cranes in the Port of Southampton were four overhead travelling cranes, installed in 1903 in sheds 34 to 36. These were followed in 1904 by one of 50-ton lifting power alongside Number 6 Trafalgar Dry Dock, which was replaced in 1951 by one of the same capacity but with increased height and radius. It was not until the opening of the Ocean Dock (the former White Star Dock) in 1911, however, that the first electric quayside cranes were introduced – a batch of eight by Stothert & Pitt of Bath, who have supplied most of the cranes for Southampton Docks. From that year, electric quayside and dry-dock cranes gradually superseded hydraulic cranes.

➤ Quayside cranes in action in the Eastern Docks in May 1961. (Gwilym Davies/Britton Collection)

The lifting power of the cranes at Southampton varied in the post-war era from the gigantic 150-ton floating crane at 106ft radius to the 2-ton quayside cranes with a radius of 65ft. Until 1939, the majority of the quayside cranes were of the 2-ton and 5-ton types, but the experience gained in the Second World War demonstrated that 3-ton and 6-ton types were more suitable to the demands of the post-war era. With modernisation planning in the late 1950s and early 1960s, most deep-water berths were eventually equipped with larger cranes with an 86ft radius and a reach of 70ft.

Visitors to the 102 Berth building could see four 30cwt-capacity overhead travelling cranes, which operated on a semi-open gantry bay. They were used to transfer baled cargoes from the upper floor direct to ground-level

waiting trains. To witness the speed of them in action was remarkable. This was largely due to the piecework operation, which was extant when the author saw the cranes at work. By contrast, a visit to the Old Docks in the late 1950s/ early 1960s could reveal petrol-driven mobile 3½-ton and 5-ton cranes at work in the sheds. Long heavy freight trains would arrive from Feltham Yard with open wooden wagons hauled by former Southern Railway S15 Class steam locomotives. Their loads of railway equipment, machinery and Austin cars – export products – would be unloaded by the mobile cranes. Having completed unloading, the mobile cranes would then set to work loading up the empty wagons with all sorts of imported items from as far afield as South Africa, India, Australia or New Zealand. In the 1950s, some of these mobile cranes were steam-powered and it was exciting to watch them snort out great clouds of white steam when they struggled with lifting the heavy loads.

The most powerful cranes, apart from the 150-ton floating crane, were the two 50-ton cranes. One was located on the west side of the King George V Graving Dry Dock. It was specially designed to clear the funnels of the Cunard Queen liners. The second was located at the Trafalgar Dry Dock.

Most of the cranes in Southampton were 'level-luffing', which meant that the hoist could travel in a horizontal path over the whole distance from maximum to minimum radius. This proved extremely useful when shifting loads between ship and quay and down into the holds of ships. The speed of a hoist was quite remarkable – 200ft per minute for 2-ton and 3-ton cranes, 100ft per minute for 5-ton and 6-ton cranes.

All post-war cranes had a standard gauge on rails of 18ft, which meant that if required they could be moved to a different location. If a crane had to be transferred to another berth within Southampton Docks, the floating crane was dispatched with a tug to transport them. This was always an interesting port movement to observe. Today this is but a distant memory, as with the introduction of containerisation, the cranes have been replaced with modern efficient gantries for rapid ship to rail or lorry transfer.

The 'star of the show' amongst the cranes of Southampton was without doubt the huge floating crane, which was a familiar sight for over sixty years. It was built by Cowan Sheldon at Carlisle in 1924 and had a maximum lift of 150 tons. The crane was not self-propelled and required the use of tugs to move it around the port. Up until 1962 the floating crane was fitted with steam machinery in order to perform lifting operations. The sounds of it chuffing and the clatter of the winding gear when in operation were quite memorable. The floating crane somehow lost its character when droning diesel equipment replaced the steam boilers. In the 1930s, the crane distinguished itself on many duties including assisting with the transfer of O2 and E1 steam locomotives and rolling stock from Southampton to Medina Wharf on the Isle of Wight. During the Second World War it was secretly transferred to the Clyde on 'special duties', but made a triumphant return to Southampton in 1946. The floating crane normally resided at Berth 48, near the entrance to Ocean Dock. Sadly, it was finally dismantled for scrap in 1985 and replaced by a new 200-ton floating crane.

13

DRY DOCKS AND SHIP REPAIRS

It was always fascinating when visiting Southampton Docks to peep behind the scenes to observe what vessels were being refitted or overhauled in the dry docks, from the smallest yacht to the largest liner. Repair work of all kinds was undertaken by skilled men to ensure that Southampton could provide unrivalled first-class maintenance facilities. Here is a summary of the dry docks:

NUMBER 1 DRY DOCK

Opened in 1846. Overall length of 401ft, 378ft at floor level and 66ft wide.

NUMBER 2 DRY DOCK

Opened in 1847. Overall length of 281ft, 240ft at floor level and 50ft wide.

NUMBER 3 DRY DOCK

Opened in 1854. Overall length of 523ft, 501ft at floor level and 80ft wide.

Numbers 1, 2 and 3 dry docks were accessed via the Outer Dock. At the top end of Number 2 Dry Dock was a pumping station. All three dry docks were officially closed on 31 December 1963 for redevelopment of the Inner and Outer Docks involving filling in the dry docks for a drive-on, drive-off car ferry terminal.

NUMBER 4 DRY DOCK

Opened in 1879. Overall length of 479ft, 451ft at floor level and 55ft wide.

On 6 November 1940, this dry dock suffered bomb damage during an air raid and there are photographs and records showing an unexploded bomb on site. Number 4 Dry Dock is officially listed as closing on 28 March 1964 and was declared in the ledger as, 'to be filled in to make way for redevelopment.' However, this brick-lined structure remained perfectly intact with its wooden gates still watertight into the late 1990s, when it was decided to fill it in and build houses on top of it for the new Ocean Gate development.

NUMBER 5 PRINCE OF WALES GRAVING DOCK

Opened in 1895. Overall length of 912ft 3in, 729ft at floor level and 91ft wide.

The dock was opened on 3 August 1895 by the Prince of Wales, later King Edward VII. It was last used in the 1970s and was filled in and tarmacked over.

NUMBER 6 TRAFALGAR DOCK

Opened in 1905, but enlarged in 1913 and 1922. Overall length of 912ft, 852ft at floor level and 100ft wide.

The original Trafalgar Dock was built with 130,000cu. yd of concrete. It had steel entrance gates operated by direct-acting vertical engines. This impressive facility had stepped sides with a series of concrete linked stairways. By 1913, however, although it was the largest dry dock in the world at that time, it was not large enough to receive the newly constructed SS *Olympic*, so it was enlarged. During the rebuilding the steel gates were replaced by a sliding steel caisson, which allowed the level of water inside the dock to be maintained against a falling tide. In 1922, the Trafalgar Dock was rebuilt again to accommodate the SS *Berengaria*. A V-shaped cleft was cut into the head of the dock into which the liner's bow fitted, leaving a mere 10in between the side of the ship and the dock wall!

The Trafalgar Dock was an officially 'listed construction', but this did not prevent Associated British Ports from filling in 90 per cent of it in and covering it in tarmac. The last few feet, including the gate, remain intact. From 1924, use of the Trafalgar Dock declined as a new floating docking was brought into use, which in turn was superseded by the King George V Graving Dock.

NUMBER 7 KING GEORGE V GRAVING DOCK (1933–2005)

Opened in 1933. Overall length of 1,200ft, 1,141ft 6in at floor level and 135ft wide.

The King George V Graving Dock was situated at the extreme west end of the New Docks on the site of former tidal mudland. It was designed by F.E. Wentworth-Shields to accommodate the new transatlantic Cunard *Queen Mary* and *Queen Elizabeth* liners, and constructed by John Mowlem & Co. and Edmund Nuttall & Sons, costing more than £2 million. When opened on 26 July 1933, it was the largest dry dock in the world and retained this distinction for nearly thirty years. The design of the dock reflected the near vertical sides of the new liners. It had steep sides with three lines of cradle blocks, avoiding the need for shoring. This gigantic dry dock took two years to construct and was built almost entirely of concrete with granite dressings for the sills and caisson stops.

The 4,000-ton caisson door slid sideways into the chamber at the right of the entrance. The adjacent pump house held four pumps which could empty the dry dock of its 58 million gallons of water in just over four hours. A visit inside the pump house revealed walls lined with cream and green tiles on the lower part. A creaky wooden staircase led to a mezzanine balcony with a wooden balustrade. When a liner entered the dry dock, the four pumps would kick into action under the push button control of one engineer at a central control

◄ The *California Star* is seen undergoing her annual refit in dry dock. She was a 8,577-GRT refrigerated cargo liner, built in 1945 as *Empire Clarendon* by Harland & Wolff Ltd, Belfast, for the Ministry of War Transport. In 1947 she was sold and renamed *Tuscan Star*, then *Timaru Star* in 1948. She was sold again in 1950 and was renamed *California Star* in 1959. (Gwilym Davies/Britton Collection)

desk, pumping 78,500 gallons (350 tons) of water per minute. As the pumps throbbed rhythmically, it was possible to view proceedings on a large illuminated indicator panel which showed in diagrammatic form a full plan view of the dry dock, pumps, valves and culverts. The panel was illuminated by various colours to show the phases of the pumping operation. Green lights would light up when the valves were closed and red lights when they were open. The author can only liken this to a railway signal box where an illuminated track panel plots the progress of a train. The author's Passing visits with his father to the King George V Dock, under the pretext of 'calling in to see Uncle Fern Bussell', were always a fascinating spectacle – once seen, never forgotten.

In 2005, Associated British Ports terminated the lease for the King George V Graving Dock with the ship repairers A&P Group. The caisson gates and keel blocks were removed and the dock became a permanent 'wet dock'. In June 2006, the dock and pumping house were both granted Grade II listed building status.

FLOATING DRY DOCK (1924–1940)

This was built on the Tyne and delivered to Southampton in 1924. Its overall length was 960ft, and it was 134ft wide. The height of the side walls was 70ft. The floating dock was transferred to Portsmouth in 1940 for use by the Admiralty and renamed ADF Number 11. In 1959 it was acquired by the Rotterdam Dry Dock Co., and in 1983 was resold to new owners in Brazil. Sadly, the floating dry dock was wrecked off the coast of Spain while under tow.

Across the River Test on the New Forest shoreline facing the New Docks was Husbands' Shipyard at Cracknore Hard. This was a privately owned facility, with its own small floating dry dock where the tugs and ferries operating from Southampton and in the Solent were maintained.

The main shipbuilding and repair company in the port was John I. Thornycroft & Co. which, although established in 1864, did not relocate to Woolston until 1904 following the acquisition of the Mordey Carney shipyard. The first ship built by Thornycroft was the Tribal Class destroyer HMS *Tattar* for the Royal Navy, and it soon acquired a fine reputation for building naval destroyers. The business flourished during the First World War with further military contracts, including a construction programme for high-speed coastal motorboats, CMBs.

After the First World War, Thornycroft took on more civilian contracts and built the ferry SS *Robert Coryndon* for operation in Uganda in 1930. In 1931 the yard built the first of three fine vessels for Red Funnel, the MV *Medina*. A sister ship, the MV *Vecta*, was completed in 1938. The third vessel in the trio, the MV *Balmoral,* was not built, however, until 1949.

Following the outbreak of the Second World War, production concentrated on Ministry of Defence contracts. Under Thornycroft's direction, some 1,929 Landing Craft Assaults (LCAs) were successfully produced for D-Day. After the war, Thornycroft continued to thrive with defence and civilian contacts and maintenance refits in the local dry docks. The company merged in 1966 with Vosper & Co and was restructured to enable it to continue building warships as well as diversifying into training and support.

14

ISLE OF WIGHT FERRIES

Up until 1774 the Isle of Wight was quite isolated, but in that year, regular passenger-carrying sailing services commenced. In 1820, the service improved with regular steam sailings following the introduction of the paddle steamer *Prince of Cobourg*. Soon, two companies were vying for passengers with new steamers on the Southampton–Cowes route, namely the Isle of Wight Steam Packet Company and the Isle of Wight Royal Mail Steam Packet Company. In 1861 these companies amalgamated to form the Southampton, Isle of Wight & South of England Royal Mail Steam Packet Company, better known to us today as Red Funnel Services. In addition to the paddle-steamer service between Southampton and Cowes, a considerable amount of excursion traffic was provided for the tourists in the summer season in Isle of Wight waters from Southampton to Ryde, Portsmouth, Bournemouth, Poole and Swanage.

The pioneer paddle steamers of Red Funnel, named *Sapphire, Emerald, Ruby* and *Pearl*, were the inspiration for the company's house flag: 'Blue to mast, green to fly, red on deck and white on high.'

Other prominent paddle steamers in the Red Funnel fleet included such illustrious names as: *Balmoral, Solent Queen, Her Majesty, Princess Beatrice, Princess Helena, Prince of Wales, Queen* (later renamed *Mauretania* in 1936 and *Corfe Castle* in 1937), *Princess Mary* and *Lord Elgin*. Such names conjure up many memories. The PS *Balmoral* was the pride of the fleet; built in 1900, she was the largest Red Funnel paddle steamer and had a fantastic speed for the time of 20 knots, whereas the PS *Solent Queen*, which was twin-funnelled, had a reputation for engine trouble.

The PS *Lord Elgin* was a most interesting old craft, unique in being the last surviving cargo paddle steamer in Britain. Her engines were not the original

set, though of a similar type. Her mast, which carried a derrick, was abaft (in the stern half of the ship) and she was flush-decked. Originally, the PS *Lord Elgin* plied for the Bournemouth, Swanage & Poole S.P. Co. Ltd, but when she passed into Red Funnel's fleet she was employed latterly with cargo only, between Southampton and the Isle of Wight. Unfortunately, this historic maritime vessel (a good candidate for preservation) was broken up at Northam in 1955.

A feature of the Red Funnel's paddle steamers was their high standard of maintenance. This was noticeable to the inquisitive passenger who 'casually peeped' into the engine room of the PS *Princess Elizabeth*. In one word, it was 'exceptional', the machinery glistening – a delight to the eyes and a credit to the engineers in charge.

The Red Funnel paddle steamers, like their counterparts operating the Portsmouth–Ryde service, gave a good account of themselves in the war years. Sadly, the famous PS *Gracie Fields* (the last paddle steamer built for Red Funnel, in 1936, and named by the great lady herself) was lost in an air attack. The PS *Lorna Doone* was converted for use as a minesweeper and distinguished herself when attacked by three Dornier bombers. She shot down one, damaged a second and drove off the third. A BBC Radio programme was produced in her honour.

Red Funnel Steamers believed in emulating the habits of the chameleon and had various colours for different services and changed their colours at least three times down the years. The colours of their paddle steamers, up to the time that they obtained the purely excursion paddlers, were buff funnels, black hulls and paddle boxes with brown saloons. In 1931, Red Funnel instituted a change of colours. After a short trial of giving a black top to the funnel,

➤ The Red Funnel Isle of Wight ferry *Medina* at the Royal Pier on 1 August 1959. (Gwilym Davies/Britton Collection)

all their steamers had their funnels painted white with a black top. In 1935, another change was made: this was the red funnel with black top. It was, in fact, the colouring of the Bournemouth & South Coast Steam Packets. With it the excursion steamers, together with *Princess Elizabeth, Gracie Fields* and *Queen*, retained their white upper works and paddle boxes, but the older paddle steamers retained their brown deck houses and black paddle boxes.

Alas the day of the Southampton-based paddle steamers sailing down Southampton Water and across the Solent has disappeared into history. Today the paddle steamer is a Victorian antiquity, a tourist attraction in its own right, whose memory at Southampton is perpetuated by occasional visits from the preserved PS *Waverley*.

Red Funnel's first passenger screw vessel was the MV *Medina*, which was built locally in 1931. She was followed by the larger MV *Vecta* and MV *Balmoral*. After the Second World War, motoring on the Isle of Wight became very popular, and Red Funnel acquired, in the late 1940s, a surplus tank landing craft to convert to a roll-on, roll-off vehicle-carrying vessel. It was said that the introduction of this vessel, more than any other factor, contributed to the closure of the island railway branch lines to Ventnor West, Freshwater and Bembridge.

The Red Funnel success of the introduction of a roll-on, roll-off passenger and vehicle service was followed up in 1959 with the construction of a purpose-built passenger car ferry, the *Carisbrooke Castle*. She revolutionised travel on the Isle of Wight and could carry 45 cars and 900 passengers. Her introduction, it is said, was the nail in the coffin of the Cowes–Newport–Ryde and Shanklin–Ventnor railway lines. Today, Red Funnel dominates ferry services from the Isle of Wight, with passenger vehicle ferries and a superb high-speed service that is second to none.

Only once has Red Funnel's dominance of the Southampton to Isle of Wight ferry service been challenged. In July 1966, British Railways introduced a hovercraft service between Southampton and Cowes. It was operated under the Seaspeed banner, using two SRN6 hovercraft from Crosshouse Road (next to the Woolston Floating Bridge) to the Cowes Medina Road Terminal. During the winter of 1971–72, both hovercraft were lengthened by 10ft, with capacity increased from thirty-six to fifty-eight passengers, and named *Sea Hawk* and *Sea Eagle*. The service transferred to Hovertravel of Ryde in 1976, but hovercraft operations from Southampton ceased in 1980.

15

REMINISCENCES OF SOUTHAMPTON DOCKS

Captain John Davidson, Red Funnel:

At the Port of Southampton there were two companies which shared the duty of supplying tugs for towage work around the docks, Southampton Water and the Solent – namely the Alexandra Towing Company and the Southampton & Isle of Wight Steam Packet Co., Red Funnel. I worked for Red Funnel, who had owned and operated tugs since 1885, starting as a deckhand and working my way up to finish as captain of the tug *Hamtun II*.

Our Red Funnel tugs could be identified by their dark red funnels with a black top, whereas the Alexandra tugs had a yellow funnel with a black top divided by a broad white band and black ring. Originally, our Red Funnel tugs were based at 101 Berth in the New Docks, while the Alexandra Towing tugs were based at 46 Berth in the Ocean Dock. Latterly, however, both tug fleets were transferred to the Empress Dock.

Our competitors, the Alexandra Towing Company, were a Liverpool-based firm and first came to Southampton at the invitation of Cunard when they transferred their crack liners from Liverpool to Southampton in 1919. It would be fair to say that Red Funnel commanded the lion's share of the Southampton tug work as we had the larger fleet. In saying this, both companies' tugs worked together as a team when docking and undocking the large passenger liners. All the crews from both companies knew each other well, and although there was much banter and friendly rivalry, at the end of the day we were all mates and would help each other out.

When I started working on the Southampton tugs in 1955, the Alexandra Towing Company tug fleet consisted of *Sloyne*, *Wellington*, *Gladstone*, *Flying Kestrel*, *Brambles* and their tug tender *Romsey*. Our Red Funnel tug fleet consisted of *Canute* (skipper Captain Harold Hurst), *Clausentum* (skipper Captain Steve Pascoe), *Hamtun*, *Hector* (skipper Captain Jagger Noyce), *Neptune* (skipper Captain Follet), *Sir Bevois* (Captain George Howard), *Vulcan* (skipper Captain Little) and the tug tenders *Calshot* (skipper Captain Curley) and *Paladin* (skipper Captain Byrne). Of the Southampton-based Alexandra tugs, only their tug tender *Romsey* (skippered by Captain Ken Gregory) came close to matching the power and reliability of our Red Funnel tugs. Without wishing to sound boastful or biased, the pick of the tug fleets was Red Funnel's *Calshot*. She was robust and reliable, the master of all work. Her reputation soon spread and liner captains and docking pilots would request her assistance in preference to all other tugs. At one time or another I have worked relief duties on all the Red Funnel fleet and observed at close quarters the Alexandra tug fleet in action and can therefore honestly say that my preference of tugs was fair, and this view was supported by many of my contemporaries.

Our main tug work around the port was to tow and assist with the manoeuvring of ships in and out of the docks, moving ships from berth to berth and in and out of dry docks, day and night on a twenty-four-hour round-the-clock basis in all weathers. The 'cream on the cake' prestige jobs were moving the world's largest liners, including the Cunard Queens, SS *United States* and SS *France*, but to achieve this required years of experience. Southampton-based tug work in the heyday of the ocean liners was a highly skilled job. It was a career often started as a deckhand working your way up the promotion ladder, which you grew up with. The knack of handling large liners was only learned through experience. Added to this, in Southampton, one had to know every sand bar and hidden underwater hazard, the effects of the double tide and currents from the Itchen and Test rivers. Strong easterly high crosswinds in the winter months were another factor to consider, especially when turning a gigantic passenger liner like the SS *United States*, which had two large funnels that acted like enormous sails. These threatening winds were unpredictable, and a sudden gust could result in serious trouble for the assisting tugs.

In the good old days of the steam tugs around the Port of Southampton, it was necessary to maintain a good boiler full of steam and for this a good stoker was required. With the introduction of diesel-power tugs, the constant challenge for keeping up steam pressure disappeared. I would say, however, that our Red Funnel steam tugs had more power and could, when required, rescue a situation from potential disaster.

There were many types of tugs in the two Southampton-based fleets, including those for fire fighting and pollution control. Both companies possessed tug tenders, which were used for liners not entering the port. For instance, before the Second World War, Red Funnel would always send a tug tender, like the *Calshot* and *Paladin*, out to Cowes Roads to meet the French Line *Normandie,* which was at anchor. The passengers alighting for Southampton and the Waterloo trains would transfer from the liner to the tug tender and travel to the Royal Pier in the lounge, smoke room or bar below the main deck. It wasn't just passengers who transferred from the waiting liner to our tug tender, as the anchored liners would often hoist over motor cars, mailbags and passenger luggage. This was quite a challenge if there was an easterly gale blowing, which resulted in passengers getting wet through from spray and the tug tender uncomfortably rising and falling on the waves. On such occasions we received many complaints and the tips from the impatient first-class passengers were few!

We would sail from the Royal Pier with a tug tender down Southampton Water taking on occasions up to 100 passengers to meet a liner that was anchored off in Cowes Roads. On some trips we would be joined by VIP first-class passengers like Dorothy Lamour, Ginger Rogers, Burt Lancaster, Cary Grant and Grace Kelly. Miss Kelly tried to disguise herself with a head scarf and dark glasses to avoid glaring passengers and press photographers. We all felt sorry for her, but there was no escape on such a small craft.

The tugs at Southampton were often employed to move 'dumb' harbour equipment, such as barges, floating pontoons, dredgers and the 150-ton floating crane. This could be quite a challenging task as barges have a mind of their own. We were sent out with the floating crane to assist with the transfer of some railway carriages to be removed from Medina Wharf on the Isle of Wight to Southampton. All was going well when suddenly the wind increased just as the jib was lifting a crimson-coloured 57ft carriage. It swung back and forth like a paper toy. My big fear was that the whole lot would release and come crashing down on the deck of our *Calshot* tug. I was never so glad to return in one piece to Southampton.

On occasions the tug fleet would receive a call to drop everything and go out into the Channel beyond the Isle of Wight in response to a salvage call. We therefore always kept a good supply of extra food and drinks on board, just if the occasion required, as you never knew when that call may come. When the call did come for help, it was usually at night when the weather was bad. On occasions it was to come to the aid of a stricken liner. Just before I started with Red Funnel there was an emergency call to assist the RMS *Queen Elizabeth* which had run aground on the Brambles on 14 April 1947. Red Funnel and Alexandra Towing mustered every spare tug from Southampton to pull the *Lizzie* clear of the sandbank, but in the end it took extra help from the Admiralty tugs at Portsmouth.

The usual routine when docking an inbound ocean liner like the *Queen Mary, Queen Elizabeth* or *United States* was for the Trinity House pilot to board the incoming liner off the Nab Tower, near Bembridge on the Isle of Wight. One of our Southampton-based tugs would then proceed from the docks to rendezvous with the liner off Ryde Pier or in Cowes Roads off the Isle of Wight. This tug would then act as an escort round the turn of the Brambles and up Southampton Water. With this tug would go the docking pilot, customs and immigration and one of the liner company's officials. They would transfer from the deck of the tug to the liner via a Jacob's ladder and enter via the liner's shell door. This could be tricky when it was rough weather, but when I was in charge we never lost anyone overboard. On several occasions, however, I have known outbound liners to abandon the transfer of pilots and officials due to rough weather. In such cases they usually accompanied their charges to Le Havre or Cherbourg and returned to Southampton the following day.

Similarly, I have known of occasions when a New York Sandy Hook pilot has had to unexpectedly come across the Atlantic with a liner, because transfer had been impossible at the Ambrose Lightship off the entrance to New York.

With inbound liners, the other tugs from both companies would be assembled and waiting off Netley, like a squad of smart soldiers, ready for the approach of the liner. When we heard three short blasts from the liner's whistles, each tug would move off into a prearranged position to receive the tug's heaving lines, which were dropped from a hawser on the liner. Once the lines were secured to each tug's towing hook by a 12in manila rope we would take up the tension. When completed, each tug would signal to the liner they were in position. There were generally two tugs ahead, two on the port quarter and two under the port bow with a seventh tug ready to go onto the port quarter to assist the nose of the liner into berth. On occasions when the weather was bad, the pilot (usually Captain Jack Holt) would call for extra tugs.

For those tugs required to push, a bow fender would act as a bumper when the tug's nose was against the liner. Believe it or not, there were 1½ tons of rope and rubber in these fenders, which were handmade by the tug's crew, largely in their spare time.

We all worked as a team when working with the big liners at Southampton. Everyone knew each other and it was a matter of close co-operation acting under the instructions of the pilot on the liner. He alone was in sole charge of the liner and us as masters of the tugs. Our orders were communicated from the pilot on the bridge of the liner via the liner's whistle or via a pea whistle.

Here are the signal codes:

From the liner to the forward tugs via a pea whistle:

1 short whistle – tow ship's bow to starboard
2 short whistles – tow ship's bow to port
1 short whistle – tow ahead of ship
A succession of short whistles – cease towing forward.

From the liner to the after tugs via the ship's whistle:

1 short whistle – tow ship's stern to port
2 short whistles – tow ship's stern to starboard

1 short whistle – tow ship astern
A succession of short whistles – cease towing aft.

This system remained pretty much the norm until the introduction of the radio telephone.

When there was a very high tide it has been known for liners to enter or sail from Southampton via the Needles, on the Isle of Wight, between Yarmouth and Lymington. This would save time and, more importantly, fuel for the parent shipping line who owned the liner. This unexpected treat would be at the discretion of the liner's captain in consultation with the pilot. For instance, on Thursday 12 October 1967, the *Queen Elizabeth* sailed from Southampton on an incoming high tide. As we approached Calshot, with the tide approaching its highest point, the order suddenly came at about 7.15 a.m. from Commodore Marr on the Cunard Queen: 'Change of plan. We are going out via the Needles!' Now, Danny Marr knew the draught of the *Lizzie* and was a great opportunist. He seized the moment with a successful outcome – typical of the man. The passage of the larger liners via the Needles was always a rare occurrence and it could catch out tug crews.

My career on the Southampton tugs began after coming out of the Royal Navy when I joined Red Funnel in the post-war heyday of the ocean liners. My regular tug on which I spent most of my career as the mate was the *Hamtun I*, which was built and launched locally at the Thornycroft yard at Woolston in 1953. I joined her two years later in 1955. She was a powerful modern twin-screw steam tug with 1,500hp triple-expansion engines and was the first of a new series of Red Funnel steam tugs at the time of her launch. She was fitted with a typical Thornycroft-style funnel top, which was designed to lift smoke and smuts. The 318-ton *Hamtun* was oil-fired as against coal-fired, which took away a lot of the back-breaking sweaty work to raise steam. I guess it was like a marriage between the *Hamtun* and me, as I knew her from top to bottom and loved her dearly until she was replaced in October 1970. At times she could be awkward, but if you knew how to treat her with love and care, she responded well. The Red Funnel system for tug crews was that you remained with your regular tug for nine months per year and for the remaining three months crews were allocated 'down the river' to be based on the two standby fireboat tugs at Fawley.

After the demise of the *Hamtun*, I became a relief captain on various Red Funnel tugs, until being allocated to the new diesel-powered *Calshot II*. I had previously worked on her namesake, the now preserved steam tug *Calshot*, and I eventually retired in 1992. Mention of the original *Calshot* brings back the memory of the time when we had to bring in the *Queen Mary*. We were attached aft, and after bringing the liner into Ocean Dock the crew were keen to finish the shift and head off home. We watched as the ship safely tied up and one by one all the tugs let go and headed away, except the old *Calshot*. As time passed, our crew began to shout out: 'Hey, let go!' The language directed at the *Queen Mary* crew began to become a bit choice as hunger kicked in. I therefore scrambled up the thick rope line on the seaward side of the *Mary* until I reached the porthole where the rope originated from. I poked my head through. There before me was a chap quietly smoking while a group of the *Queen Mary's* deck hands were engrossed in a card game! 'When you are ready, will you let go of the Calshot!' In shock, with white faces, they all jumped up in great surprise with cards and money cascading to the deck. 'Sorry lads,' came the apologies from above.

The crew on the old *Calshot* were just like one big family. They were a tough breed, but we would all look after each other. On the *Calshot* there were three high and three low firebox doors stoked by two firemen. Now, on one occasion one fireman did not turn up and we had to sail with just one fireman, Tossy Toswell. This stoker always took pride in keeping up full steam pressure. Unfortunately, on this occasion, while using the dart tool to rack through the fire to keep it 'white and bright', poor Tossy burned himself. After an initial yelp, he just continued firing for twenty-five minutes. He muttered not one complaint and made sure that the needle on the boiler gauge was showing a full head of steam.

As I climbed the promotion ladder to a master of a tug, many fellow captains offered helpful advice. Amongst them was Captain Steve Pascoe, who had joined Red Funnel during the war in 1942. Steve had a vast experience and was involved in towing the Mulberry Harbour sections from Southampton following the Normandy Invasion of 1944. He also was in command of the *Clausentum* in 1947 and helped free the Cunard *Queen Elizabeth* off the mud at Calshot. Steve watched over me and offered many helpful tips, especially when assisting the large Atlantic liners. In 1961, Steve became the master of the *Thorness* and ended his career on the *Calshot II*. Steve would always praise and encourage fellow captains and crew. His vast experience was put to good use in the early 1970s when he was called on to assist with a number of oil pollution dispersal operations. It therefore came as no surprise that he was nominated for an award in recognition of his outstanding service at Southampton, and he was presented with the MBE by the Queen Mother at Buckingham Palace.

In the 1960s, the artist Terence Cuneo joined us for a week on our tug and merrily sketched away with his pencil and charcoal on an artist's pad. There were portrait sketches of the crew, liners, cranes, quayside buildings, etc. The artist was preparing material for studio publications and ideas for a large oil-on-canvas painting commission. Cuneo also attempted to paint from life a portrait of a Cunard Queen in Ocean Dock, but he soon gave up as we were unstable through bobbing around too much. A few weeks after he had finished and returned home, we received a parcel through the post of some Cuneo sketches. Amusingly, he had sketched a mouse captain at the wheel of our tug! A closer look at the sketches revealed that there was a hidden mouse in each sketch, which was his apparently Terence Cuneo's trademark artistic signature.

Both the Red Funnel and Alexandra Towing Company tugs have been extensively featured in various television documentaries and BFI (British Film Institute) British Transport films. One Red Funnel tug, *Paladin*, became a film star in its own right. In 1958, a film company hired the tug and crew to star alongside the actor Peter Sellers in the film *The Mouse That Roared*. The *Paladin* took a central role in this film as the boat which made an imaginary voyage across the Atlantic carrying the army of the Grand Duchy of Fenwick to invade New York. For the crew of the tug this was a very lucrative affair, as many scenes had to be repeated with retakes – some as many as seven times! Each time this happened, the crew would receive additional payments, and so by the end of the week they were all rich and smiling.

Regrets? Only being away from wife and family for so long, working antisocial hours when the children were growing up, but if I could turn the clock back I would step back aboard my Red Funnel tug tomorrow as long as I could bring an old Cunard Queen into Ocean Dock.

Joe Webb:

Like many Southampton families, there was a tradition for the Webb family to either go to sea or work in the docks. My father, Harry Webb, worked in the docks, but I initially chose to go to sea with Cunard on the RMS *Queen Elizabeth*. My brother, Bernard, also joined Cunard on the *Mauretania* and later *Queen Mary*, beginning as a bellboy before promotion to a steward. After a few years at sea, I decided that long voyages away from my young wife and family were not for me, so I decided to seek a job working in the docks.

It was tough and physically demanding working on the dockside in all weathers manhandling huge amounts of cargo or manhandling passenger luggage and trunks, but, living at Millbrook, it was convenient and my first pay packet was £3 10s. There was an army of stevedores and crane operators employed in the docks in the days before the age of the container. In those days, cargo was transported in crates or cases and sometimes in heavy bales on and off the ships. Anything of value or of great weight was lifted by crane from the deep holds of the ship directly onto the quayside. From here the loads were transferred straight into waiting railway wagons or transported into the stores and warehouses.

We would often spend a few days unloading a ship, but today the job of unloading containers is completed in hours. I vividly recall the South African fruit ships would require their holds open for a number of hours to disperse toxic gases before we could commence unloading. This was especially true with citrus fruits like oranges. The job brought us into contact with all parts of the world in the days before television and computers came on the scene. Royal Mail ships would arrive from South America loaded with chilled and frozen meats. Nearer home, Channel Island ferries would come into port with fresh tomatoes or Jersey potatoes. Who needed a geography lesson? One of the more unusual seasonal loads arrived every Christmas in the form of boxes of mistletoe and turkeys from France. Going in the opposite direction for export out of Southampton to all parts of the old British Empire were Massey Ferguson tractors and agricultural equipment and cars from Coventry, railway equipment from Eastleigh, Swindon and Crewe Works. This was in the days when Britain proudly built things and exported to the world. In those days it all came through Southampton, which was a hive of activity.

Working in the docks could be very dangerous, particularly at night in wet, windy, icy or foggy conditions. I recall we were unloading one of the Union-Castle boats at 103 Berth when a crate full of fresh South African oranges blew free of the unloading crane jib and came tumbling down. The oranges cascaded and bounced everywhere over the quayside. Unfortunately, once the oranges were scattered on the concrete the authorities refused to accept them for the wholesale markets. There was no alternative other than to scoop them up and take the fresh juicy Cape oranges home as a free treat for the family.

◄ Former Hollywood screen icon Princess Grace (Kelly) of Monaco and Prince Rainier sailed on the SS *United States* from Southampton to New York in September 1956. As they passed through the Ocean Terminal they were witnessed having a very public disagreement about the name of their future baby! (Associated Press)

In the 1950s, bananas began to become increasingly popular. They were usually brought by Geest and Fyffes ships to 24/25 Berth or 40 Berth. At first the bananas were unloaded in cases from ship to shore by hand. This was back-breaking hard work and sometimes very dangerous too! On one occasion I was unloading a banana boat and I suddenly felt something painful on my leg. At first my leg was numb and then it began to throb. Looking carefully I spotted two puncture marks and I suspected the worst. Immediately I ran from the quayside jumped on my motorcycle and drove at 60mph straight through the dock gates, past the bemused dock police, straight to the hospital. Dumping my motorcycle, I ran through into the Casualty Department and showed my wounds to a doctor. He confirmed the worst. It was a snake bite. 'A mamba?' I meekly asked. He smiled and replied, 'No, if it had, you would have been dead by now.'

On another occasion I spotted a dark ink-black beetle climbing out of a banana crate. I carefully placed it into a tobacco tin and put it in my pocket. On returning home, I decided to play safe and gas the beetle before sliding the tin in the fridge. That night I took the tin with my trophy down to the local pub to show my fellow dock workers what I had caught. I opened the tin up and assumed that the beetle was well and truly dead. A crowd of strong and brave Southampton stevedores gathered around closely in eager anticipation and curiosity. As I opened the lid, the warmth of the bar resurrected the beetle and it sprung out like a jack-in-the-box with frog-like legs! The bemused stevedores scattered in blind panic and with it their pints of Strongs Best Bitter, leaving me to pay up 11s for their replacement.

You never quite knew what to expect when unloading bananas at Southampton Docks. One morning the stevedores gingerly opened the hold of a Geest ship. As they peered down through the gloomy darkness into the depths of the ship, before them were revealed sixteen white pairs of eyes. It was impossible to distinguish any faces or other features in the darkness. Work stopped immediately and the dock police were summoned to the scene. Beams of white light from the torches shone down to illuminate sixteen West Indian stowaways. The customs and immigration people came down and interviewed these illegal visitors. Unusually, they decided to issue them with overalls, boots and gloves and set them to work unloading as payment for their passage!

The age of manually unloading bananas fast began to disappear in the late fifties and early sixties, as at 24/25 Berth four new overhead gantries were built to assist with unloading. Manual dock labour was

▲ From 1960, for the next seventeen years, Union-Castle Line's *Windsor Castle* made one hundred and twenty-four voyages between Southampton and Cape Town. After the liner had unloaded her passengers, the author recalls venturing onboard with his father to sample fresh South African fruit – courtesy of an uncle who was assisting with cargo transfer! (John Wiltshire)

reduced with the introduction of endless chains with canvas pockets which were lowered into the holds of ships. The canvas pockets carried the bananas into the sheds ready for transfer onto waiting steam-hauled trains. This mechanised system allowed up to 14,000 stems of bananas to be unloaded an hour.

A lot of the work in Southampton Docks was routine, but we could be working under pressure at times to meet sailing times. One such routine pressure came from the Union-Castle line schedule. Every Thursday at 4 p.m., a Union-Castle ship would sail from Southampton bound for Cape Town in South Africa. At the same time a Union-Castle ship would sail from Cape Town to Southampton. The Union-Castle boats were mostly named with the suffix *Castle* in their names. In the 1950s the Castle boats were well known for their lavender coloured hulls with red funnels topped in black. They would arrive on a Friday at 35 Berth, and were unloaded for transfer to 102 or 104 Berth in the New Docks where they sailed from on the Thursday at 4 p.m. I would regularly see the Castle-named ships *Arundel*, *Stirling*, *Athlone*, *Capetown*, *Pretoria*, *Edinburgh*, *Pendennis*, *Transvaal* and *Windsor*. Just a little way down the New Docks at 106 Berth was the home of the P&O-Orient Line. Scenes at the quayside at 106 Berth were always full of mixed emotions. There were passengers full of excitement and expectation waving and cheering as they sailed off on a cruise to some far-off tropical location. This was contrasted by the sad sight of families emigrating from these shores to Australia, sometimes with tears in their eyes, leaving behind loved ones and families and not knowing what to expect on arrival down under. The P&O cruise liners always had distinctive funnels, such as the *Arcadia, Himalaya, Chusan, Strathmore, Orsova, Oronsay* and *Orcades*. The P&O fleet were later joined by the more unconventional-looking *Canberra* and *Oriana*.

A familiar sight when working a little further down the New Docks at 109 Berth was the troop ships *Devonshire, Oxfordshire, Empire Fowey, Empire Ken, Empire Windrush* and *Empire Orwell*. Bibby Line and Orient Line managed these troopships for the Ministry of Transport. The troopships were instantly recognisable by a dark-blue coloured band painted on their white hulls. Companies of 500–600 troops would frequently arrive at Southampton Docks on a special troop train. Alternatively, they would come alongside 109 Berth and pile out of Bedford army lorries. It could take up to a few days to load the troops and their supplies before they sailed for far-off tropical army bases such as Aden, Malaya or Hong Kong. On occasions a military band was quayside to serenade departing or returning regiments. When inbound troopships returned with squaddies, you could always see they were smiling, happy and laughing – some often whistling or cheering. The officers would come down the gangplank and the first words they would often say were: 'Great to be back home in Blighty.'

Just about everyone who worked as a stevedore in the docks acquired a nickname and it was possible to know someone throughout their entire working lives by just their dock nickname. These nicknames originated either from the way they dressed, their habits, speech, looks or physique. My nickname was Garth, as my colleagues sarcastically referred to my slim and wiry physique and said I required muscles. Names bestowed on each other and often by best mates in the Southampton Docks were accepted without malice and often brought you closer together, and indeed your real name was forgotten. There were the usual Joners, Greenies, Clarkies, Smiffies, Scotties, Taffies, Paddies, Snowies, Lofties, Tinies, Titches etc. I would often meet fellow dock workers in the town while shopping and we would exchange pleasantries referring to each other by our dock nicknames, much to the amusement of our wives and families. My wife, Jean, would look puzzled and ask: 'Strange name. Who was that?'

Amongst the stevedore nicknames I recall in my Southampton stevedore 'A to Z' were:

All Electric – not a spark in him

Arab – who went around muttering, 'Well, well'

Battleship – who was always after a sub

Bernie the Bolt – only way to keep him on the job

Billy Deodorant – a real stinker

Bitter Bob – could not wait to finish his shift and head to the pub for a pint of bitter

Black Bob – water baby

Blanket Brothers – never took off their overcoats, even in the heat of summer

Bread and Jam – named after his favourite sandwiches

Buck – relation to Rogers

Bulkhead – always saying: 'Once we get the bulkhead in we're laughing.'

Buzz – a busy bee

Captain Beaky – so-named because of his large nose

Celery Legs – I think he used to bleach his legs

Chicken Lips – who refused to wear his dentures

Cinderella – he always had to be away from the night shift by midnight

Clive of India – always claimed he met Gandhi off the boat from India

Clock 'Em In – foreman in the New Docks who was keen on punctuality

CoCo – a total clown at the best of times

Corrugated – named after his wavy hair

Cowboy – every inch a John Wayne-type docker

Dead & Buried – a docker who would not work without others

Double Wrap – who always complained about feeling the cold

Dr Mo – a chap who had a cure for every dock worker's aches and pains

Drop a Goolie – who was always getting things wrong

Effy – who swore non-stop

Electric Eel – so-named for being a live wire

Fireman's Jacket – who always wore a coat with polished shiny metallic buttons

Flipper – who always stopped short of swearing, 'Flip this. Flip that...'

Gannet – always hungry

Gee Gee – said he owned a horse

Gravel Voice John – spoke with a frog in the throat

Hand on the Hips – who always stood with his hand on his hips during rest breaks

Harry call me John – who always introduced himself with: 'My name is Harry, but call me John.'

Horace Hotplate – who was always first in the canteen breakfast queue

Jelly Legs – who walked along the quayside with a wobble

Kybosh – who always put the kybosh on jobs when unloading liners in the New Docks

Man without a Cause – whose constant complaint was 'I think it's 'cause the foreman don't like me.'

Marconi – never stopped telling his mates what he had listened to on the radio

Mario Lanza – known for his singing

Mick the Mint – always in after eight

No socks – because he never wore any socks

Oncer – Joe Clark, who often requested a tip from the first-class transatlantic passengers at Ocean Terminal: 'Give me a oncer'

Pincher – who frequently asked 'Is there a copper on the gate?'

Reverend Day – who always preached to others

Ronnie the Rat – who was scared of the semi-pet cats that were in the docks

Shifting Sands – was always dodging from one shift to another

Sinbad – often came to work looking like a wreck

Stop the Job – a shop steward, named after his favourite saying

Sweep – who often swept chimneys of fellow dock workers on his way to the docks

Tea Pot – who always cheered up everyone by saying 'Get the teapot on.'

Ted the Balloon – who constantly told his men not to let him down

Ten Past – who often arrived ten minutes late for work

The Beast of Belsen – a rather aggressive individual who remained caged up operating a crane

The Count – a noble-looking dock worker

The Ghost – appeared at the quayside when a ship docked, but was never there when work was underway

The Lawyer – always asking 'What do you want to know?'

The Long-Distance Runner – who would often arrive late for work saying he missed his bus

The Mafia – someone who was often heard to comment, 'Let's keep it to ourselves'

The Sheriff – who would ask, 'What's the hold up for, lads?'

Tick Tock – who would reprimand anyone tempted to help themselves to a Cape orange or apple when unloading at 102 Berth, saying, 'Put it back.'

Waiting Time – who would enquire of the foreman 'How much waiting time we got for this ship?'

Whoa Whoa – who would hold up his arms to approaching USA dock steam locomotives and shout, 'Whoa! Whoa, driver!'

I wonder if the traditions of Southampton Docks continue in this modern age of the container?

➤ With a long train of South African loaded fruit in tow, USA 0-6-0 tank No 30066 crawls through the streets past Southampton Town Quay. A number of these USA tanks built by Alco, Baldwin and Porter companies in the USA, to USA Army Transport Corps design, worked in the UK during the Second World War. At the end of hostilities, fourteen of them were purchased by the Southern Railway specifically for shunting at Southampton Docks. (Alan A. Jarvis)

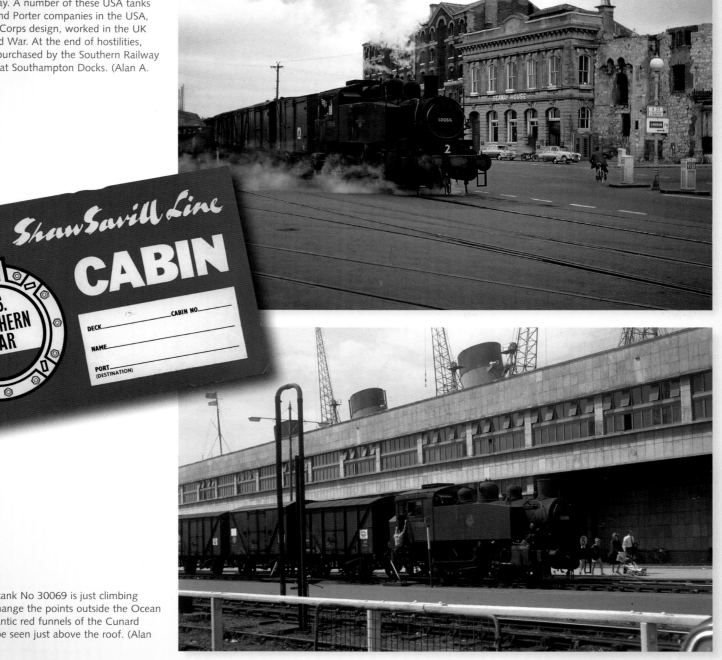

Shaw Savill Line

CABIN

S.S. NORTHERN STAR

DECK_____ CABIN NO._____

NAME_____

PORT_____
(DESTINATION)

➤ The fireman of USA tank No 30069 is just climbing down from his cab to change the points outside the Ocean Terminal. The three gigantic red funnels of the Cunard RMS *Queen Mary* can be seen just above the roof. (Alan Sainty Collection)

⌃ For this special occasion, Merchant Navy Class No 35004 *Cunard White Star* was provided by Nine Elms locomotive shed in London to haul the *Mauretania* Regent Refinery opening special train. A circular smoke box name board with the name *Regent* was attached. The special train is pictured at Ocean Terminal station in Southampton shortly after arrival. The engine has been polished to perfection for the occasion on 27 October 1964. (Valero Energy/Texaco)

➤ This aerial view of the Eastern Docks was taken in 1960 and shows an empty Ocean Dock. There is some activity in the New Docks, however, as behind the Union-Castle Line ship in 101 Berth, there are tugs preparing to assist with the sailing of a P&O vessel from 102/103 Berth. (Esso/Britton Collection)

➤ Continuing the theme of the outward-bound Union-Castle Line, this photograph shows *Capetown Castle* on 26 August 1954 heading away past the Eastern Docks escorted by the Red Funnel *MV Balmoral* departing with the Isle of Wight service to Cowes. It was in February 1965 that £100,000 of gold ingots went missing from the *Capetown Castle* while in Southampton! After a thorough police search of the ship, with dogs and a forensic search team, the majority of the missing stolen gold was discovered cemented into the mast house with quick-drying cement and the remainder hidden in a children's sandbox on the deck. (Pursey Short/Britton Collection)

◄ Viewed from the landward end of Ocean Dock in June 1966, we have a clearer view of RMS *Queen Mary* and *Carmania*. The dockyard was silent and empty apart from three other shipping enthusiasts and the distinguished artist Terence Cuneo, who can be seen busy sketching the *Queen Mary* on the left of this picture. (Norman Roberts/Britton Collection)

◄ The sheer size of the Cunard RMS *Queen Mary* was impressive from any angle, especially from water level. She is pictured here in Ocean Dock. (Britton Collection)

This sequence of slides shows the sad final departure of the Cunard RMS *Queen Mary* from Southampton on Saturday 16 September 1967. Surrounded by escorting local tugs and proudly flying her paying-off pennant, she sails regally down Southampton Water into history. (George Garwood, World Ship Society)

︿ This very rare, evocative, immediate post-war early-morning scene was shot in colour at Ocean Dock. Local fishermen in their demob greatcoats are seen busily casting their lines, while on the opposite side of the dock is the veteran Cunard RMS *Samaria*. (Pursey Short/Britton Collection)

▲ This is the Ocean Terminal building in Ocean Dock. The building was completed in 1950, and on 31 July 1950 was officially opened by the British Prime Minister, the Rt Hon. Clement Atlee MP. This stunning building was built in art deco style and was 1,297ft in length. It was called the Southampton Ocean Terminal and was designed to cater for the transatlantic liner services such as the famous Cunard Queens and other famous transatlantic liners of that time. There were sumptuous reception halls for the passengers on the first-floor level and on the ground floor space for ships stores and freight. Also, there was a railway platform to cater for boat trains from London Waterloo bringing passengers to the liners. The RMS *Queen Elizabeth* was the first ship to use the new terminal. The passenger reception halls were linked to the ships via three pairs of power-operated telescopic gangways. Conveyor belts could also take supplies and goods on and off the ships quickly and efficiently. Sadly, by the early 1980s, use of the magnificent Ocean Terminal had declined. In 1983 it was tragically demolished, somehow evading a preservation order. (Gwilym Davies/Britton Collection)

▲ A pallet of luggage is gently craned from the quayside at Ocean Dock into the hold of the Cunard RMS *Queen Elizabeth* in 1953. (Arthur Oakman/Britton Collection)

➤ Stevedores sit around doing their newspaper crosswords and enjoying a cigarette awaiting the order to cast off all mooring lines prior to the sailing of the *Queen Elizabeth* from Ocean Dock. The building on the left is the magnificent Ocean Terminal. (Norman Roberts/Britton Collection)

◄ The deck officers of the *Queen Elizabeth* and Her Majesty's customs officers are on the quayside at Ocean Dock in this 1953 picture. They are inspecting cars prior to loading in the forward hatch. (Arthur Oakman/ Britton Collection)

UNITED STATES LINES

CABIN C
STATE

◄ The Red Funnel tug *Canute* has steam up and is ready to haul the *Queen Elizabeth* out of Ocean Terminal. (Pursey Short/Britton Collection)

➤ Commodore Geoffrey Marr, Senior First Officer Gossett and Senior Southampton Pilot Jack Holt, standing on the port-side wing of the *Queen Elizabeth*, gaze back into Ocean Dock as the ship completes her swing manoeuvre. They are watched by a good crowd standing on the upper observation floor of the Ocean Terminal building. (Austen Harrison/World Ship Society)

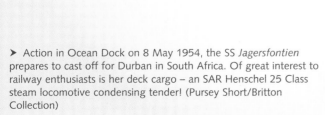

➤ Action in Ocean Dock on 8 May 1954, the SS *Jagersfontien* prepares to cast off for Durban in South Africa. Of great interest to railway enthusiasts is her deck cargo – an SAR Henschel 25 Class steam locomotive condensing tender! (Pursey Short/Britton Collection)

⌃ This very rare sequence of pictures was taken on the 30 May 1953 and shows the arrival of the United States Line *America.* She is escorted by the Alexander Towing Company tugs *Wellington* and *Flying Kestrel*, together with Red Funnel tug tender *Calshot,* and tug *Canute.* The whole of the proceedings is observed by Asian sailors from another vessel tied up in Ocean Dock. They appear to be more concerned about what fish they may catch. Once tied up, we see *America* framed ibetween two vessels, one of which is the *Esso Plymouth* on the left. Finally, *America* is dramatically pictured illuminated at night awaiting her next duty. (Pursey Short/Britton Collection)

◄ Elder Dempster Line's *Aureol* first started sailing regularly from Southampton to Africa in April 1972. Seen at first hand, one had to agree that she was a truly eye-catching ship, reflecting an almost large yacht design. (John Wiltshire)

▼ Now renamed *Carmania*, the refitted former RMS *Saxonia*, sets out from Ocean Dock on a cruise. However, she is once again repainted, this time in white livery. (John Goss)

The veteran Welsh photographer Gwilym Davies must have been a great fan of the French Line CGT SS *France*, for he took over 150 colour slides recording the arrival and docking of the liner at Ocean Dock in Southampton. Here are just a few. (Gwilym Davies/Britton Collection)

➤ This superb dramatic night shot of the SS *France* in the Ocean Dock was used by the publicity department of the French Line on a poster. The original slide has remained hidden for almost fifty years. The name of the liner, *France*, is illuminated on the top deck for all to see. (Gwilym Davies/Britton Collection)

▼ The former *France* was purchased by Christian Klosters of Norwegian Caribbean Cruise Lines, who had her rebuilt and renamed *Norway*. Here is a tribute to this famous liner, taken when she visited Southampton in 1982. (Norman Roberts/Britton Collection)

⌃ Congestion in the Docks. Memories of the Seamen's Strike in June 1966, taken from an Isle of Wight Red Funnel ferry. (David Peters)

⌃ An unusual visitor to Southampton Docks in 1958 was the railway steamer *Duke of Lancaster*. She was about to undergo a refit in the local dry docks. The *Duke of Lancaster* escaped scrapping and is currently beached near Mostyn Docks in north-east Wales where there are plans to transform the vessel into an art gallery. (Gwilym Davies/Britton Collection)

⌃ The Cable & Wireless ship *Stanley Angwin*, a frequent visitor to Southampton Docks. Photographers would often train their lenses on the more glamorous ocean liners at Southampton Docks and overlook the routine workhorses of the port, but not so photographer John Goss. (John Goss)

▲ Local yachtsmen who had their yachts based on the River Hamble would often sail around the docks with visitors to view the liners. Here a group of sightseers sail to see the Cunard RMS *Mauretania* in 1958. (Gwilym Davies/Britton Collection)

▲ This spectacular picture taken in 1960 shows the departure of the Cunard RMS *Mauretania* assisted by Alexandra Towing Company and Red Funnel tugs. It was taken by the legendary American Braun Brothers during a visit from New York. The documentation with the original slide is very interesting, revealing that it has quite a history – it was submitted for a photographic competition in the USA, which it won, and was made available to a souvenir company in the United Kingdom. (Braun Brothers)

▲ A favourite in Southampton Docks was the TSS *Golfito*, which was a passenger-carrying banana boat launched in 1949 belonging to the Fyffes Line. She was 8,687 tons and 448ft long. The *Golfito* had three passenger decks with cabins for ninety-four first-class passengers, public rooms and an open deck space centred between four large refrigerated cargo holds. (Gwilym Davies/Britton Collection)

Three views of the car ferry *Patricia*. From 1964, with the arrival of a number of car-ferry companies in the port, it was decided to redevelop the Inner and Outer Docks to accommodate them. The Inner Dock was filled in and the area used for car parking. The entrance to the Outer Dock was widened, and new buildings were constructed, including a timber-arched passenger reception hall adjacent to Berths 2 and 3 in the Outer Dock. The new complex was opened in July 1967 by Princess Alexandra and named in her honour. This photograph shows the Swedish Lloyd's ferry *Patricia* at Berth 3. She made her first sailing to Bilbao on 5 April 1967. (John Goss)

▲ The Elders & Fyffes TSS *Changuinola* was a regular visitor to Southampton in the 1960s. She was also a vessel that was infrequently photographed, but was typical of the Southampton scene. (John Goss)

▲ A rare glimpse of the troopship *Asturias*, which is being refuelled. The *Asturias* made twenty-four trips to Australia between 1946 and 1952, carrying more than 30,000 migrants. Most of them came under an assisted-passage scheme through which adults travelled to Australia for just £10 and children travelled free. The voyage followed the Mediterranean Sea via Malta, the Suez Canal, Aden in Yemen, Bombay in India and Karachi in Pakistan before arriving in Fremantle, in Western Australia, which took between five and six weeks. In 1953, the Ministry of Defence converted the *Asturias* into a troopship to repatriate British troops involved in the Korean War. She was withdrawn from service in 1957 and sailed to Faslane in Scotland for scrapping. (Gwilym Davies/Britton Collection)

The SS *Tetela* was a cargo and occasional passenger ship that belonged to the banana-importing firm Elders & Fyffes, a wholly owned subsidiary of the United Fruit Company. In this sequence of pictures, photographer Gwilym Davies recorded her arrival and docking assisted by the Alexandra Towing Company tug *Formby*. (Gwilym Davies/Britton Collection)

⌃ Pictured in the Eastern Docks on the Itchen estuary is the *Steel Fabricator* cargo ship, which is being unloaded. (Gwilym Davies/Britton Collection)

⌃ A procession of lorries have crossed the railway lines and are seen weaving their way into the docks with exports passing the Ocean Buffet. (Norman Roberts/Britton Collection)

⌃ HMS *Gurkha* was launched at the Thornycroft Woolston shipyard on 11 July 1960. Here we see the hull being fitted out on this Tribal Class frigate ready for commissioning by the Royal Navy on 13 February 1963. (Gwilym Davies/Britton Collection)

⌃ A splendid view of the New Docks taken from 37 Berth in the Eastern Docks, showing the Glasgow-registered *Pegu* tied up at 40 Berth. (John Goss)

⌃ The Ben Line's *Ben Macdhui* is pictured docked at 41 Berth in the Eastern Docks. Ben Line was a Scottish shipping company, based at Leith, which exported from Southampton to the Far East. (Gwilym Davies/Britton Collection)

⌃ With sister Union-Castle ships pictured in the New Docks, the *Rothersay Castle* is seen being unloaded by quayside cranes at 41 Berth in the Eastern Docks. (John Goss)

⌃ Four Red Funnel tugs turn the Royal Mail Line *Andes*. The RMS *Andes* was built by Harland & Wolff, Belfast, and was due to be delivered to Royal Mail Lines in 1939, but due to the outbreak of war, she instead became a troopship. In 1947, she was at last delivered to her owners and commenced work on the company's services from the UK to South America. In the 1950s and 1960s she was a very popular and frequent visitor to Southampton. (Gwilym Davies/Britton Collection)

P&O

NAME
(IN BLOCK LETTERS)

S.S.

CABIN OR BERTH NO.

KINDLY PLACE ONE LABEL
ON EACH END OF THE PACKAGE

CABIN

◄ A view of the Chandris Line cruise ship *Britanis*, which entered service in October 1931 as the SS *Montery* for Matson Lines. During the Second World War she was used as a troopship and was operated as a fast transport ship capable of sailing independently. (Britton Collection)

▼ The smoky and rusting *Seaborne Alpha* provides an evocative scene at Cracknore Hard on the shores of the New Forest opposite the New Docks. (Gwilym Davies/Britton Collection)

⌃ This picture was taken from the very end of Hythe Pier, and behind the Hythe ferry can be seen the three funnels of the Cunard RMS *Queen Mary* resting in Ocean Dock. (Gwilym Davies/Britton Collection)

⌃ The troopship *Nevasa* can be seen in the New Docks on 19 August 1961 in this view taken from Cracknore Hard. (Gwilym Davies/Britton Collection)

⌃ This view of a rusting wreck taken at Cracknore Hard contrasts well with the pristine Union-Castle Line *Edinburgh Castle,* which is seen berthed in the New Docks opposite in April 1961. (Gwilym Davies/Britton Collection)

➤ A study of the funnel of a British Railway Channel Island ferry at Southampton Docks in May 1961. (Gwilym Davies/Britton Collection)

⌃ Pictured in Dry Dock Number 6 Trafalgar Dock is the Union-Castle Line *Rowallan Castle*. She was a product of Harland and Wolff in Belfast, built in 1943 for the Union-Castle Line. (Pursey Short/Britton Collection)

⌃ An unusual picture, taken from sea level, of a State Marine Lines' cargo ship in 42 Berth in the Eastern Docks. (Gwilym Davies/Britton Collection)

⌃ This fascinating picture shows the British Railways steamer *Deal* undergoing refit in the Number 5 Prince of Wales Graving Dock on 25 March 1961. This dry dock had an overall length of 912ft 3in (729ft at floor level) and was 91ft wide. The dock was opened on 3 August 1895 by the Prince of Wales, the future King Edward VII. It was last used in the 1970s and was then filled in and tarmacked over. (Gwilym Davies/Britton Collection)

➤ Lying beside the dry docks is a fascinating collection of new propellers, which are all carefully labelled and ready to be fitted to ships entering the dry docks for refit. (Norman Roberts/ Britton Collection)

➤ This is possibly now a unique and historic picture of the Southampton Dry Docks 'T-Urn' taken in August 1957. This boiler was used to supply tar and pitch for corking wooden decks with rope. It was an essential tool, which has now disappeared with the introduction of modern steel decking. (Pursey Short/Britton Collection)

This sequence of pictures shows the first lorry load of infill rubble being dropped into the dry docks. These sad pictures signalled the end of the historic Number 6 Trafalgar Dock. (Britton Collection)

⌃ The Town Quay dock cranes are pictured in action moving loads of timber. This area is now a huge car park – almost unrecognisable. (Norman Roberts/Britton Collection)

⌃ The 1925-built 150-ton floating crane is seen on the move being towed by two Red Funnel tugs. (Norman Roberts/Britton Collection)

⌃ Taken from the deck of the Cunard *Queen Elizabeth 2*, the 1925-built 150-ton floating crane is seen on the move being towed by Red Funnel and Alexandra Towing Company tugs. (Graham Cocks/Britton Collection)

⋏ Left: The Shaw Savill Line *Southern Cross* is underway down Southampton Water sailing from Southampton for a voyage around the world. (Britton Collection); Right: The Dorset photographer Tom Hedges took this superb picture of the MS *Achille Lauro* sailing down Southampton Water. She was a cruise ship based in Naples, Italy. Built between 1939 and 1947 as MS *Willem Ruys*, a passenger liner for the Rotterdamsche Lloyd, she is best known for having been hijacked by members of the Palestine Liberation Front in 1985. Sadly, in 1994, the ship caught fire and sank in the Indian Ocean off Somalia. (Tom Hedges)

◄ The Cunard RMS *Queen Elizabeth* makes a dramatic smoky entrance in Southampton Water as she is slowly turned into Ocean Dock by the Red Funnel and Alexandra Towing Company tugs. (Gwilym Davies/Britton Collection)

➤ The Cunard *Carmania* is seen slowly passing Calshot at the southern end of Southampton Water. (Gwilym Davies/Britton Collection)

➤ The British India Line educational cruise ship SS *Uganda* is seen passing Calshot returning from the Mediterranean. (Gwilym Davies/ Britton Collection)

◀ The Esso Oil Refinery from the air on Saturday 14 August 1954. (Pursey Short/Britton Collection)

◀ An oil tanker is unloading her vital cargo at the Esso Oil Refinery at Fawley in the mid-1960s. Today, the massive supertankers that visit Fawley have to be seen to be believed. (John Cox)

These very rare colour pictures show the Princess flying boat on 13 September 1953 flying over Calshot and Southampton Water. This is one of the very few occasions that this massive flying boat flew. (Pursey Short/Britton Collection)

▲ The Princess flying boats are pictured here under construction at Saunders Roe at Cowes on the Isle of Wight in 1952. (Pursey Short/Britton Collection)

▲ The 140-ton Calshot Spit Lightship marked the entrance to Southampton Water. After decommissioning in 1978 she was sold to be cocooned in concrete in the new Ocean Village housing complex. (John Wiltshire)

▲ This view, taken in 1958, shows the lightship on station. (Norman Roberts/Britton Collection)

▲ It was always of great interest to sail past the Calshot Spit Lightship – this view was taken from the Red Funnel ferry *Carisbrooke Castle*. (Graham Cocks/Britton Collection)

➤ The Cunard RMS *Queen Elizabeth* is framed by the houses and the two shores of Southampton Waters as she heads from the Solent towards Spithead. (David Peters)

➤ The Cunard RMS *Queen Elizabeth* is now clear of the navigation hazards. Heading towards Ryde, past Cowes, she is seen just ahead a small oil tanker. (David Peters)

⌃ This 1958 view of the *Queen Elizabeth* was taken from above the tree line at Egypt Point, Cowes, on the Isle of Wight. We can see the flame at the Fawley Oil Refinery and witness just how close inshore the massive Cunard liner is during her complex 'zigzag' manoeuvre around the Calshot sandbanks and in to Cowes Roads. (Britton Collection

◄ Peeping through a porthole on the Cunard RMS *Queen Elizabeth* in 1953, as we pass Norris Castle on the Isle of Wight. (Arthur Oakman/Britton Collection)

▲ The Russian cruise liner SS *Batory* is seen passing Egypt Point at Cowes on the Isle of Wight heading towards Southampton. (David Peters)

◀ Next stop Le Havre. The 66,348-ton SS *France* has just left Southampton and is passing Cowes Station where O2 tank number 26 *Whitwell* is running around her train. (David Peters)

◀ Ocean liners were not the only beautiful ships to sail from Southampton. Here we see the sailing ship *Sangress* off the Isle of Wight on 17 August 1954. (Pursey Short/Britton Collection)

⋏ *Calshot* was a tug tender built in 1929 by John I. Thornycroft & Co, and completed in 1930 for the Red Funnel Line. She was put into service tendering the various liners that stopped either in the Solent or Southampton Water, which saved them the time and expense of docking just to take up or set down a few passengers. She was also used during the summer season for tourist use in the excursion fleet. Calshot is pictured here in Ocean Dock on 31 July 1959. (Gwilym Davies/Britton Collection)

⋏ An action shot of the 'stern swing' of the Cunard RMS *Queen Elizabeth* with the Red Funnel tug tender giving 100 per cent. (Braun Brothers/Britton Collection)

⋎ The Red Funnel tug *Hamtun* is seen in action assisting the Shaw Savill Line *Northern Star*, which has just arrived in Southampton from Australia. (Braun Brothers/Britton Collection)

▲ The Red Funnel tug *Calshot II* in March 1966. (Norman Roberts/Britton Collection)

⌃ An unidentified Red Funnel tug heads into dock passing some keen fishermen in September 1958. (Gwilym Davies/Britton Collection)

⌃ The Alexandra Towing Company tug tender *Flying Breeze* and tug *Ower* await their next turn of duty in 1966. (A.E. Bennett)

⌃ The powerful Alexandra Towing Company tug tender *Romsey* awaits her next turn of duty. (Gwilym Davies/Britton Collection)

➤ Car congestion along the quayside of Berth 45 as a bevy of Alexandra Towing Company tugs and the service ship *Twyford* await further duties. (Norman Roberts/Britton Collection)

▼ Berth 45 was the home of the Alexandra Towing Company tug fleet. In the foreground are the tugs *Gladstone* and *North Loch*. (Norman Roberts/Britton Collection)

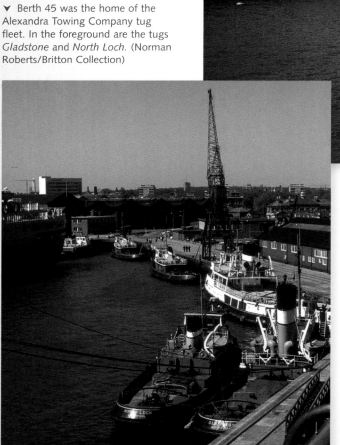

➤ Alexandra Towing Company tugs *Northern Isle* and *Gladstone* rest in Ocean Dock in March 1966. (Norman Roberts/Britton Collection)

⌄ Left: The Alexandra Towing Company tugs *Brambles* and *Canada* await their next turn of duty on 1 September 1958. (Gwilym Davies/Britton Collection); Right: Alexandra Towing Company tugs *Northern Isle* and *Gladstone* rest in Ocean Dock in the company of the RMS *Andes* in April 1966. (Gwilym Davies/Britton Collection)

◄ A picture of a rare visit to Southampton Royal Pier by the British Railways Portsmouth–Isle of Wight paddle steamer PS *Ryde*. (John Wiltshire)

◄ The stoker on the British Railways paddle steamer is seen using the dart tool to rake through the firebox and build up steam at the Royal Pier, Southampton. (John Goss)

UNITED STATES LINES

FIRST CLASS STATEROOM

➤ Stirring times at Southampton Royal Pier as the Red Funnel paddle steamer *Princess Elizabeth* sets sail for Cowes, Isle of Wight. (Braun Brothers/Britton Collection)

➤ The highly polished engine room on the Red Funnel paddle steamer *Princess Elizabeth* was a place that was always spotlessly clean and a great pleasure to visit. (Braun Brothers/Britton Collection)

▲ The British Railways paddle steamer PS *Sandown* is returning to Ryde from the Royal Pier with day trippers from Portsmouth. (Britton Collection)

▲ Inside the British Railways paddle steamer, PS *Sandown*, there was a commemorative plaque recording the paddle steamer's war service situated below her builder's plate. (Britton Collection)

▲ The Red Funnel Isle of Wight ferry *Vecta* at the Royal Pier on 25 March 1961. (Gwilym Davies/Britton Collection)

▲ The Red Funnel Isle of Wight car ferry and former landing craft *Norris Castle* is seen unloading vehicles at the Royal Pier on 31 July 1959. (Gwilym Davies/Britton Collection)

▲ The Red Funnel Isle of Wight ferry *Medina* sails for Cowes at the Royal Pier on 1 August 1959. (Gwilym Davies/Britton Collection)

▲ The new Red Funnel roll-on, roll-off car ferry *Carisbrooke Castle* sails from the Royal Pier at Southampton bound for Cowes on 3 September 1960. She is seen passing the Lykes Line *Shirley Lykes* and Royal Fleet Auxiliary ship *Plum Leaf*. (Gwilym Davies/Britton Collection)

◄ The Red Funnel roll-on, roll-off car ferry *Norris Castle* approaches the Royal Pier at Southampton from Cowes. (John Wiltshire)

◄ The Itchen floating bridge chain ferry at Woolston in March 1966. (Norman Roberts/Britton Collection)

▲ Mrs Jean Webb of Bitterne Park drives her egg delivery van onto the Itchen floating bridge chain ferry at Woolston. (John Wiltshire)

◄ The Itchen floating bridge chain ferry at Woolston with the under-construction Royal Navy Frigate HMS *Gurkha* at the John I. Thornycroft ship yard on 3 September 1960. (Gwilym Davies/Britton Collection)

▼ The harbour master's launch in full flight in September 1958. (Norman Roberts/Britton Collection)